OF ORIGIN
Le Miocène
Châteauneuf-du-Pa
ROCKS AND GRAVEL
2012 DRY CREEK VALLEY
ACKE
MORAINE
SWARTLAND WINE OF ORIGI
2011
ELS
MYLONITE WHITE
Table Wine
The Silurian
Chardonnay
Pinot Noir
OUS
VIGNON
OBSIDIAN
WAIHEKE ISLAND
Slate
RD 89800 PREHY FRANCE
OGIEN
Orthogneiss
FRANCE
CADET
SYNCLINE
HUNTER RIVER
PORPHYRY
TECTONIC
RED WINE BLEND
ANITE
12.5%VOL
750ML
MIS EN BOUTEILLE PAR JEAN-MARC BROCARD 89800 PREHY FRANCE
CHARDONNAY EN SOL
PORTLANDIEN
Trilobite Farms
ca
RICA
2015
FELDSPAR
Pinot
TUFAIE
TITE
Quartzite
2016
Xenolith

Taste the **Limestone,** *Smell the* **Slate**

Published 2025 by Académie du Vin Library Ltd
academieduvinlibrary.com
Founders: Steven Spurrier and Simon McMurtrie

Publisher: Hermione Ireland
Editorial Director: Susan Keevil
Design: James Pople
Index: Hilary Bird

ISBN: 978-1-917084-62-8

Printed and bound in Latvia

Taste the **Limestone,** *Smell the* **Slate**

A geologist wanders through the world of wine

Alex Maltman

Académie du Vin Library

Contents

Foreword

Charles Curtis MW

One of the most fascinating aspects of the study of wine is its varied nature. An acquaintance with many topics helps a wine lover understand their passion: geography, topography, horticulture, history, chemistry, olfaction, gustation, meteorology and geology are all involved. Perhaps this last empire of knowledge is most hermetic to outsiders. If an intuitive grasp of different wine regions is relatively straightforward, with their hills and valleys, passing storms and different types of vines, grasping the intricacies of geology has always struck me as more complex, starting with the fact that the scale of time involved transcends the human. It was my own personal struggle to understand such things that led me to discover Alex Maltman.

From the beginning, my career in wine has been focused on sales and marketing. Trained as a chef, I have always had a passion for tasting. Unfortunately, the science underpinning the wine was some way down the list. As Professor Maltman notes, geology is widely bantered about in wine circles. To sharpen my knowledge of it, I set out to learn the basics. Among the first works I read was his book *Vineyards, Rocks and Soils* (2018), a foundational volume that takes the reader from electrons, atoms and ions to minerals and rocks – igneous, sedimentary and metamorphic. Thence, we are led to the folding of the rocks into anticline and syncline and to the actions of faults, scarps and grabens before discussing erosion and alluvium. These are the fundamental concepts of geology as they affect the world's wine-growing regions, and Maltman's skill in explaining them was an enormous service to to a wine-soaked general readership.

Thus, it was with great delight that I learned that Professor Maltman would publish a new work, *Taste the Limestone, Smell the Slate*. The present work uses the building blocks of geological knowledge, so skilfully laid out in his first work, to analyse and dissect the complex impact of geology on the taste of wine – and to debunk that impact when necessary. As Maltman explains in his Preface, most of *Taste the Limestone* is composed in whole or in part of articles published in *The World of Fine Wine* magazine. The first chapter sets the tone, outlining with clarity the fundamental geological concepts as they apply to the world of wine while refraining from overstating their importance to the final taste of wine.

Although geology is fundamental to wine, Professor Maltman is quick to deny that specific aromas or flavours are derived directly from the earth. The soil and rocks anchor the vines firmly in place, and the water-holding properties of the soil provide much-needed moisture, yet they have neither aroma nor flavour of their own. Maltman goes on to explain that the nutrients that are present in geological minerals are unavailable unless they become dissolved or through the action of fungi called 'mycorrhizae', which help the vines absorb them as solids. Then, in Chapter 2, he takes time to explain the mysteries of photosynthesis, including its role in the nourishment of a grapevine.

With Chapter 3, 'Musings on Minerals and Metaphors', one can almost imagine Professor Maltman saying, 'Minerality? Bah, humbug!', analysing several fashionable tasting note descriptors such as 'slate' and 'graphite' to explain why it is impossible to 'taste the limestone, smell the slate'. The term 'Metaphors' in the chapter title provides a clue. He does not dismiss the current wave of tasting notes so much as take them to task for a lack of precision. I have long explained to my students that the tasting terms used to describe wine are better understood as similes (or analogies or metaphors) than as an objective description of aromas. We all know that Pinot Noir doesn't smell like cherries, nor Cabernet Sauvignon like blackcurrants or Sauvignon Blanc like gooseberries. Similarly, Chablis does not smell like oyster shells, Mosel Riesling does not smell like slate, and Left Bank Bordeaux does not smell like graphite. But the 'truth' of the tasting note endures if it conveys to the reader a sense of what he or she may expect from the aroma of a given wine. Ultimately, our words fail us; to rely on them is to see through a glass darkly and succumb to the allegory of the cave. To 'leave the cave', in vinous terms, for the bright light of day, can be done only by tasting, not by reading a description written by someone else of their own tasting experience.

Each chapter in the book has its theme, many examining specific rocks or soils. One chapter discusses solid, immutable granite, and another explores the shape-shifting shale, slate and schist, which form more of a continuum than three distinct elements. Clay is given great respect, and the reader realizes the complexity of a material they once considered uncomplicated. In the same way, the differences between tuff, tufa and tuffeau are explained – they are not at all the same – and the connection between flint and flintiness is discussed (and discarded). Limestone comes in for special treatment thanks to its 'charisma'. If its charismatic claims are debatable, the unique properties of limestone are abundantly evident. Because of its composition (calcium carbonate), it dissolves in water and gives alkaline soils. Although it is especially favoured as a soil for viticulture, the disadvantage of its elevated pH is that certain nutrients are absorbed by the vine only with difficulty, which can lead to chlorosis and problems setting fruit.

Another thread within the chapters is to tie the geological underpinnings of wine to a sense of place. There is a disquisition on the Loire Valley from the Côtes du Forez all the way to Muscadet, and shorter inquiries into the Kimmeridgian soils of Chablis, the albariza of Jerez, Rutherford Dust (there isn't any), the terra rossa of Coonawarra, and Hawke's Bay's Gimblett Gravels – renowned, but so devoid of organic matter that the viticulture is nearly carried out, as with hydroponic vegetables, without benefit of soil.

The magic of *Taste the Limestone* is that it clearly and succinctly explains the fundamentals of geology to a wine-curious public in an engaging way that enriches their knowledge while dispelling false myths and deepening their appreciation of all the varied aspects of the world of wine. As a bonus, the book is lavishly illustrated with photographs, many taken by Professor Maltman himself. What makes the book really captivating, in my experience, however, is the fact that he is able at every turn to bring his topic to life by using anecdotes and stories that resonate with his readers, making this book a delight to read.

Preface

Professor Alex Maltman

Twenty years ago, I published a little article in a geological magazine that talked about geology, naturally, and also about wine (together with whisky and beer). Wine and geology was a pretty unusual combination back then: you rarely saw the two words together. But I was staggered by the response. Apparently the publishers had put out a press release about the article that had somehow got into the global media system. As a result, I soon had enquiries from all around the world – all in striking contrast to the feedback I'd been receiving from the rarefied, technical, academic papers I had been producing until then.

It seemed I had found a niche. Interest in the newly fashionable idea that vineyard geology can affect the wine in our glasses was just beginning to take off and it seemed that as a seasoned wine lover who grows vines and makes wine – and a trained geologist – I was able to bring some insight. So over the following years I wrote further articles, now for wine magazines, some short and snappy, others – like those in the splendid *The World of Fine Wine* publication – more extensive and thorough. The pieces received encouraging plaudits: it seems they were serving a useful purpose and were valued.

But magazine articles are, by their very nature, ephemeral. Useful or not, they tend to get put aside or be discarded, especially over 20 years and more. Much of their content may still be very relevant but, well, it's in *magazines*. Books are so much more enduring. Hence I had the idea that an anthology of some of the magazine articles I had written could be a useful contribution to wine writings generally. And here it is.

The core of the collection is a dozen articles published over the years in *The World of Fine Wine*. Some are just as they appeared there, apart from the layout and the illustrations; others have been modified in order to incorporate passages from shorter articles published elsewhere. I have also edited certain sections in order to minimize duplication and to give some continuity between the chapters. The remaining chapters were written specifically for this volume, to round out the content.

Chapter 1 involves passages from several places but centres on a 2014 piece in *The World of Fine Wine* entitled 'Between a Rock and a Hard Place'. The present title is from the *Decanter* magazine article 'On Rocky Ground' (which was retitled for the magazine's online version as 'Busting Terroir Myths', wholly inappropriately as the word 'terroir' isn't even mentioned). This opening chapter introduces some of the basics of vineyard geology to set the scene for what follows. Also, it indicates why the current vogue for a direct influence of geology on wine taste seems to me exaggerated – a theme in a number of the chapters that follow.

Chapter 2 relates the historical context of the soil–wine connection, a story that led to the discovery of photosynthesis. It was written for this volume, as was Chapter 3. The following 11 chapters were originally published in *The World of Fine Wine*. I wrote the last three chapters for this book, with the final one veering into the related world of spirits and beer. It's based on a little article in a geological magazine I wrote 20 years ago…

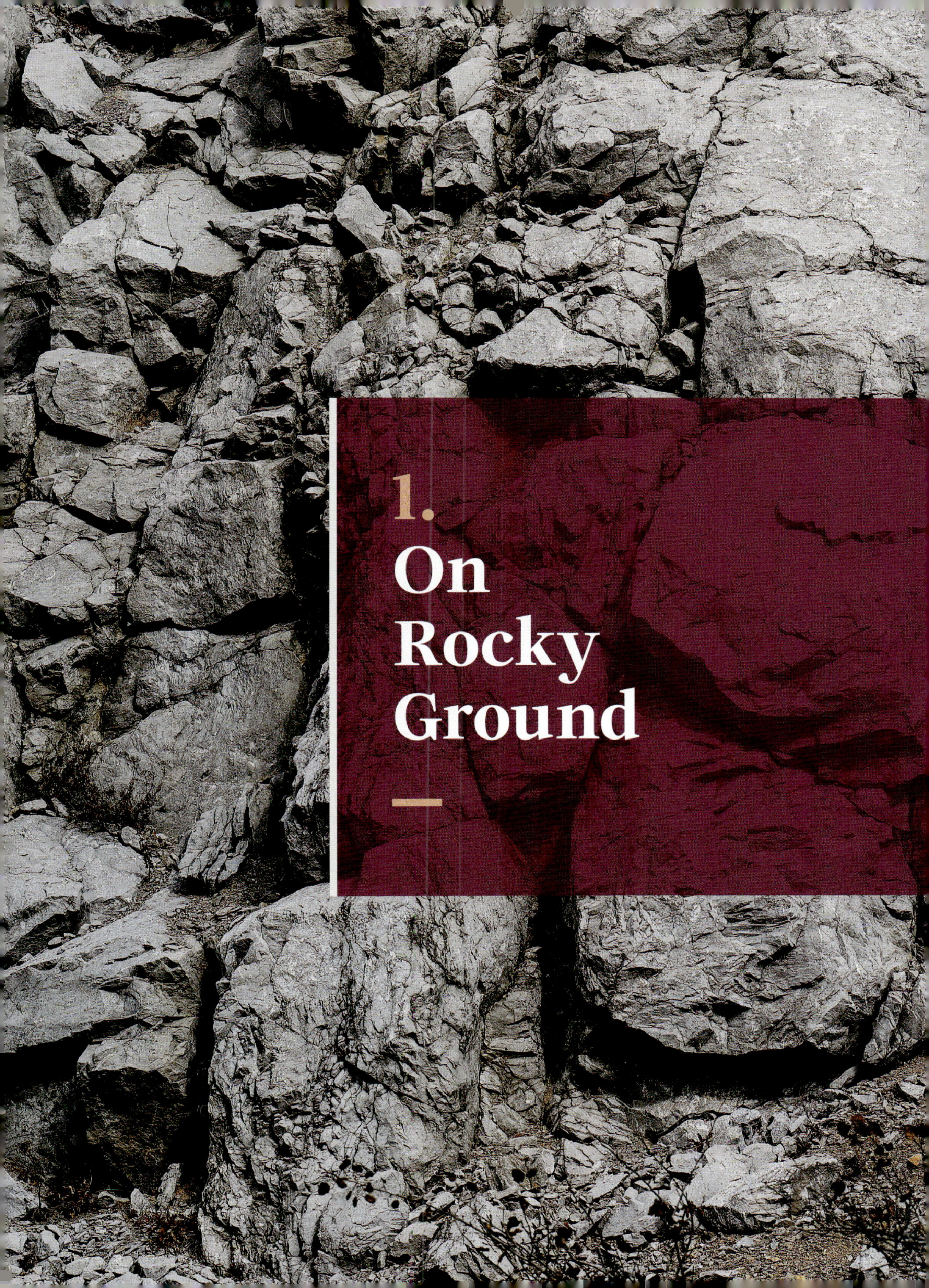

1. On Rocky Ground

On Rocky Ground

What strikes you on first seeing a vineyard? Perhaps the vines themselves? Your eye may be caught by random scatterings of gnarly old bush vines or by the military neatness of rows of vines, leafy and trimmed in summer but in winter gaunt and skeletal against their trellising. But possibly more striking might be the land itself – the geology, or at least manifestations of it. For the vineyard may be a vast affair, stretching away across a flat plain, or just a tiny huddle of vines clinging to a vertiginous slope. Or anywhere in between those two extremes: it all depends on the bedrock geology.

The soil may have an eye-catching colour or be astonishingly stony, consisting of little more than rock debris. This, too, depends on the geology. So it's no surprise that geology is so often mentioned in descriptions of vineyards: it's an intrinsic part of the rich landscape of the world of wine – hence I'm writing about it here!

A link between wine and the land has been understood and treasured for thousands of years. It even survived the discovery of photosynthesis – the fact that vines and wine are not made from matter drawn from the ground but almost wholly of carbon, oxygen and hydrogen derived from water and the air (*see* Chapter 2). And geology remains central both to appreciating a vineyard and to dealing with its practicalities. The grower will want to know about their soil: it will influence their choice of rootstocks (Chapter 15) and their vine management. How the grapes ripen will, indirectly, affect the finished wine.

But in recent decades claims about the role of geology in wine have risen to new heights. And I should be jumping for joy about it. For years I've taught, researched and generally enthused about the unsung role of geology in everyday life, and now my subject is making headlines in the world of wine! 'Soil, not grapes, is the latest must-know when choosing a wine,' *Bloomberg* tells me, for example. There are suddenly restaurants with wine lists organized not by grape, wine style or country of origin, but by vineyard geology.

One book urges drinkers to choose their wines by 'looking at the source: the ground in which it grows'. There's a consortium of growers from such diverse places as Saint-Chinian, Alsace, Corsica and Valais that claims commonality of its members' wines simply because their vines are growing on schist (Chapter 7). The same could be said about the very fashionable idea of (so-called) volcanic wines.

So why am I not full of joy about this new excitement? Well, because as a scientist I have to follow the evidence and – unfortunately – this leads me to query this new pre-eminence of vineyard geology for wine. Two things bother me. On the one hand, nothing is ever said about how these connections come about, what the geology actually does, how a particular rock brings something special to the wine in your glass. And on the other hand, our present scientific understanding makes it difficult to see how such connections might happen.

Of course, all plants are influenced by the site and soil where they grow – every farmer or gardener knows that – but only with wine is the importance of the soil so trumpeted. Does the taste of a carrot hoicked straight out of the ground reflect the soil? I've never seen that claimed. Only with wine, despite the extensive manipulation and lengthy processing involved in its production, are the connections taken this far. But, of course, wine is special.

So how might this connection work? Just how important is the vineyard geology for the finished wine? Hmmm. We need to take a closer look at the vineyard ground and how it interacts with a growing vine.

It's elementary

First, we need to know what the ground is made of. Fundamentally, like all matter, it's made of chemical elements. The remarkable thing here is that although we know of nearly a hundred different elements in nature, the composition of the outermost part of the Earth is dominated by just eight of them, namely oxygen, silicon, aluminium, iron, magnesium, calcium, potassium and sodium. Preponderant among these are oxygen, at no less than 46 percent, and silicon, at 28 percent. So, there are a lot of those two things in most vineyards!

Of course, these elements don't exist as separate entities. After all, oxygen is a gas and a vineyard isn't full of bubbles and gas balloons! Rather, the elements are combined with each other to form rigid amalgamations, and we call these natural compounds 'minerals'.

When the elements combine, a tough bonding develops between them, involving the very electrons they contain. This not only gives the compound rigidity but it locks the elements into the mineral. We shall see shortly that minerals form the structural skeleton of soils and this means that those constituent elements are firmly bonded within that framework. In other words, unless special things happen, those elements will not be accessible by vine roots.

A few elements can exist independently in the Earth – gold is the great example – though we still call them minerals. But most natural minerals, not surprisingly from the numbers given above, involve silicon and oxygen, usually with other elements in attendance. We call them silicate minerals. Thus, the vast majority of minerals making up the ground of vineyards are silicates. A notable exception is the combination of oxygen and carbon that makes carbonate minerals, and especially calcium carbonate, the mineral calcite that forms limestone (*see* Chapter 11). But the other minerals you may have heard of in vineyard descriptions such as quartz, feldspar, the clay minerals, mica: they are all silicates.

We know of many hundreds of different minerals, but here's another surprising thing. In nature we find over and over again the same few of them occurring together in roughly similar proportions, and bonded together to form strong, solid masses. The name for these rigid aggregates of minerals? Rocks. It's why we think of the Earth as a rocky planet. So, incidentally, we can see from this (notwithstanding remarks in some wine writings) that rocks are not the same as minerals.

On the rocks

Scientists like to classify things – arrange them into categories – as a first step to understanding them. But rocks have always been problematic things to classify. They don't fall naturally into boxes like, for instance, animals. An elephant is either an elephant or it isn't: in general rocks aren't like that. The obvious approach to classification would be based on which particular minerals are making the rock, but this turns out to be wholly unhelpful in practice.

The system that was eventually devised, oddly enough by the same person who joked about the quality of Cognac depending on bedrock geology (Chapter 17), is a very curious one indeed. It is based on how the rock formed, hence igneous (once molten), sedimentary (deposited as sediment), and metamorphic (changed). Well-known examples of each category are granite and basalt (igneous), sandstone and limestone (sedimentary) and slate and schist (metamorphic). (Though there was a wayward student who answered an exam question on the subject by declaring that the three main kinds of rock are classic, punk and hard.)

It's a curious scheme because with very few exceptions no one has ever witnessed a rock form or is ever likely to. Consequently, naming a rock on the basis of its origin involves a major subjective inference, and there are plenty of examples of geologists disagreeing on what to call a particular rock. Moreover, the resulting single rock name can cover a wide variety of appearances and properties.

Thus, simply stating that a particular vineyard is founded on, say, schist or marl is to disregard all the subtleties and variations that can be encompassed in a single rock name. And some of the divisions between rock types are arbitrary and some are gradational. For example, some descriptions of the rocky soils of Priorat in northeastern Spain talk of slate, others of schist. Neither is wrong; it's just that the rocks fall into a 'grey area' between the two rock categories.

We call the outermost surface of the solid Earth bedrock. All vineyards are founded on this intact, rigid assemblage of minerals. And it's the differing toughness of the bedrock in different places and how it interacts with erosion that determines the form of the land. On such things the very shape and altitude of a vineyard depend, so in this way geology is fundamental, albeit a bit academic.

In places such as craggy hillsides or river banks the bedrock may be visible, protruding through the overlying material in outcrops. Scattered around may be fragments – broken from the bedrock or brought there from further afield, perhaps by rivers. Sometimes we see in mountainous vineyards, such as at Elqui (Chile) or Pritchard Hill in the Napa Valley (California), just the tops of enormous boulders. Despite their size they are nevertheless fragments; they are loose, detached from the bedrock.

Rock into soil

Any bare rock at the Earth's surface is continually under attack by weathering. Be it a rocky cliff, a stone cathedral or a tombstone, there will always be chemical reactions between its surface and the atmosphere, just like iron reacting with moisture and oxygen in the air to form rust. William Wordsworth put the effects of weathering far more picturesquely: 'Elementary particles crumbling down, overspread…with an intermixture of colours, like the compound hues of a dove's neck.'

Rock is also attacked by physical processes such as abrasion by wind and water, and the daily expansion of bedrock exposed to the sun followed by contraction at night, sometimes accompanied by ice growing in crevices. These weathering processes act to disintegrate the rock, to reduce it to detritus. At the same time, microbes and organic debris might settle on the rock surfaces, and specialized organisms such as lichens and mosses may start to grow directly on them. On dying these biological materials decay to form tiny amounts of the rotted biological material we call humus.

Further organic material might be drifting in. And sooner or later, tough little seedlings might sprout from the humus and grow up through the detritus. In turn they will die and decompose, thus adding to the accumulation of organic matter in the moist rock debris. In other words, a soil has formed.

The water and the humus are key. The Moon, for instance, is covered in rock detritus but is without water or humus – it has no life, it has no soil. Thus, soil typically consists of a physical framework provided by the geological detritus, with spaces between the rock and mineral particles filled with some combination of water, air and a greater or lesser amount of humus.

If we dig down into a field or garden, normally we see first the humus-rich topsoil and then underneath increasingly rocky material – subsoil – in which little can grow. Sooner or later, we hit bedrock. In much agriculture the crops grow solely in the humus-rich soil at the top and what lies below is of little relevance – but vineyards are different. The distinction between rock and soil is unusually blurred, because vines can thrive in thin, humus-poor, exceedingly stony soils, roots digging deep into bedrock fissures. That's why we talk of 'vineyard geology'.

But some humus has to be involved (*see* Chapter 15), even if it is sparse and hidden below the surface stones. The apparent absence of soil in the rocky vineyards of places such as Châteauneuf-du-Pape and the Middle Mosel is legendary – how can the vines grow in that? – but the root activity is out of sight, below the ground surface in moist, humus-bearing material.

PREVIOUS PAGES: The Rippon vineyards at Lake Wanaka, Central Otago, New Zealand. The form of this striking landscape is determined by glacial erosion interacting with the bedrock geology.

OPPOSITE: Close-up (field of view about 10 cm (four inches)) showing how rocks are composed of constituent geological minerals. This example, from the island of Otrøy, Romsdal Province, Norway, is largely made of red garnet, dark amphibole and white feldspar.

Ground work

The soil is vital to grapevines, first because, obviously, they don't move around but are anchored to one place – in the soil. Second, it supplies water to the vine. And how it does this is crucial for grape quality. Research has consistently demonstrated that soils suitable for a vineyard have to provide decent drainage while in one way or another storing sufficient water for dry periods. The water is pivotal to how grapes swell and ripen, and therefore indirectly to the resulting wine.

Water stores heat well, so whether the soil is dry or wet affects its temperature. For instance, some growers irrigate in times of approaching frost, because wetting the soil allows it to take in more warmth during the daytime and then release it during the night. Clayey soils store warmth better than sandy soils because of their higher water content, and hence are slower to warm in the spring but then cool more slowly. The results of this are well known: hindered flowering for vines on clay soils, but potentially an extended ripening season.

Closely linked with the soil water content is the strength of the soil. This affects how well the soil resists erosion and allows roots to penetrate, and its ease of tillage. Wherever the soil's strength is exceeded, it will start to give and, for example, a vineyard terrace will founder. In parts of Champagne's Marne Valley, the inability of the sub-surface limestone to drain percolating water adequately is leading to landslips and extensive damage to some of the vineyards. Strength also governs how the soils respond to machines crossing them and whether the land can be sculpted, say to form drainage ditches or terraces.

To some extent, pale-coloured vineyard soils can benefit the vines by reflecting sunlight to enhance grape ripening. For instance, some growers in the Leitheberg DAC area of Burgenland, Austria, believe that sunlight reflected from their pale limestone soils leads to leaner and more elegant wines, while the darker slates yield fuller wines. The converse effect is a dark soil absorbing the sun's warmth and re-radiating it at night. For instance, dark rocks are amassed in vineyard walls at Hungary's Somló, presumably to give some re-radiation effect.

There has been excitement in scientific circles in recent years about the possible importance of microbiology in the vineyard, and new technologies are revealing distinct fungal and bacterial communities at different sites. What effect these have for wine taste is at present unclear but seeing as the kingdom of fungi includes such organisms as the beneficial mould *Botrytis* and the inconvenient yeast *Brettanomyces*, it could be very important indeed. However, perhaps because all this is effectively invisible and certainly rather technical, lacking the apparent charisma of geology, such things are ignored in most wine publicity.

Food for life

Although for centuries the belief was that vines were composed of matter drawn from the soil, a series of scientific advances revealed that it wasn't really like that at all (*see* Chapter 2). Photosynthesis was discovered, a process superficially even more magical. As a result, today, at least 10 Nobel Prizes later, we know pretty well how plants carry out their growth. It turns out that vines are very largely made not from the soil but from oxygen, hydrogen and carbon derived from water and the air, everything being driven by sunlight.

Even so, in order for the photosynthetic and other organic processes utilizing these elements to work, relatively tiny amounts of other elements have to be involved: nutrients. Most of these originally came from the ground and consequently are often called nutrient minerals (Chapter 3). Although they are there in the ground, as we have seen they are locked in the structure of the geological minerals. Except by the action of certain specialized fungi (micorrhyzae), the vine roots cannot access these nutrients directly: to be absorbed, the elements have to be dissolved in the soil water.

That comes about through weathering of the geological minerals and, in particular, to their being reorganized as various clay minerals (Chapter 9) and indirectly to give humus. If those materials release some of their constituent elements into solution, they become potential nutrient minerals. Then, if conditions are right, these dissolved elements may be physically transported through films of water between the soil pores towards the vine roots. And as Chapter 3 explains, they may or may not be taken up by the vine and transmitted through to the finished wine.

The pH (acidity) of the soil is relevant here, as besides affecting the soil structure and microbiology, it influences nutrient absorption. Rocks made of silicate minerals, such as granite, sandstone and slate, tend to give slightly acid soils, with a pH somewhat less than 7. Calcareous rocks, made of calcium carbonate, tend to be more alkaline, even exceeding 7. Chapter 10 recalls the forgotten scientific hero who showed how all this governs nutrient availability.

I love the idea of old vines, with their gnarly appearance and rich heritage. I can easily imagine that wines from them taste superior. And perhaps they do, bearing in mind the proven pedigree and naturally restricted vigour of old vines. (Analogous, maybe, to ourselves? Losing vigour with age, but deepening in wisdom?) However, the notion that their deep roots are bringing special things to the vines and hence extra nuances to the wine has no basis. Deep roots may be impressive, but they will have very largely grown in response to the need to tap into any supplementary water available deep down. The bedrock at those depths will be generally intact apart from the fissures in which water may reside. By definition, there will have been little weathering and hence there will be little availability of nutrients or anything else.

But there are limits

In all the ways outlined above, in principle the vineyard soil can be crucial to the growing vine. But why 'in principle'? Because so important are some of these soil properties that if they are not optimal a grower will try to improve them. Thus, in a way, the vineyard geology is merely providing a starting point before being to some extent overridden.

Earth-moving has long been carried out at vineyards, to a greater or lesser extent. In olden days, down-slipped soil was carted back up a vineyard slope; terraces were built by hand. These days, it's normal for a new vineyard to involve major earth-moving machinery. Impenetrable soil layers or tough bedrock is attacked by machines known as rippers, to improve vine-root penetration.

Wine writers love to enthuse about the colour of the soil. Thus (my italics), one book on French vineyards tells us that the *diorites* at Brouilly 'absorb sunlight and later give off the heat to the vines, like a miniature radiator', that the *limestones* at Sancerre 'heat up slowly in the sun, and after dusk radiate their stored-up heat back to the vines', that at Faugères 'one of the great advantages of the *schist* is that it soaks up solar heat during the daytime and radiates it back to the vines after sundown'. Elsewhere I read that: '*slate* retains daytime heat to act as a night storage heater'; that '*granite* is especially heat retentive. In evenings following warm and sunny days it continues to ripen grapes even after the sun has gone down'; that '*sandstone* is famous for retaining heat'; that '*gravel* affords excellent heat retention' and so on. It seems everything does it!

But in any case, the effect will only arise if the soil is bare, and it's highly fashionable to have cover crops, weeds, grass or something growing between the vine rows. Also, the amount of re-radiated heat falls off very rapidly with distance above the soil, so the grapes have to be trained close to the ground. (And there is a school of thought that finer grapes are produced where nighttime temperatures are markedly cooler than during the day.)

Now, that all-important water. Regarding drainage, many different kinds of geological materials are suitable – gravels in Bordeaux, for example, granite soils in the Northern Rhône, chalk in Champagne – and if there are problems it's *de rigueur* to install drains. Some of the great châteaux of France's Pomerol appellation have had artificial drains for centuries to help ameliorate its clayey soil and it's said that some have recently been augmented with electric pumps.

The fact is, for the majority of the world's vineyards, at least some degree of water management is essential. And these days, vineyard sensors can advise growers how to manage the irrigation with great precision, and so help achieve the required outcomes. Provided, that is, water is available. In these days of climate change that can be an increasing challenge. Indeed, an important limiting factor in some of the world's developing vineyard areas, such Bío Bío in Chile and the Yakima Valley in Washington, is the restricted access to water.

Irrigation water is commonly brought from some distance, such that the chemistry it brings to the vine roots may well be quite different from that of the indigenous soil (as is the case with New Zealand's Gimblett Gravels, *see* Chapter 12). And the chemical aspects of a vineyard can change through such things as treated fence posts, agrochemicals and airborne pollution. In other words, the nutrients available in the soil don't necessarily reflect the natural vineyard geology. Then, as Chapter 3 elaborates, the uptake and transmission of those nutrients into the vine is a complex affair. It all has to be closely monitored by the grower – and modified as needed.

Geology and wine taste?

So, moving from the vineyard to the wine in your glass, how relevant is the geology here? Well, a very great deal, current fashion would have it, as I illustrated at the beginning of this article. Usually there's some loose relating of wine taste to the geology, as in 'this parcel of vines is from volcanic vineyards which lend a ferrous note'. But occasionally the claims are quite explicit, as in 'the vines sip on a cocktail of minerals in the vineyard soil, for us to taste in our wineglass', and some even suggest that the rocks themselves are making it through to the wine, as in 'the weathered Devonian slate is right there in your glass'. I think the most extravagant example I've seen of this explicitness is: 'The roots just seem to suck up something from the soils and squirt it straight into one's wine glass.'

In the Pfalz (Germany) section of the current listings of a much-respected UK wine merchant I read that: 'Wachenheimer Rieslings are often pungent and spicy, no doubt due to volcanic soils.' Well, I'm afraid I have doubts. Putting aside the fact that most of the Wachenheim vineyard area does not have volcanic soils, I wonder how does this connection work? How do the volcanic soils do it? Let's be clear, the 'volcanic soils' didn't come out of a volcano. They have derived through time, just like other soils, from the disintegration of the parent rock, which is what erupted, albeit 50 million years ago. The parent rock (basalt) and the soils are made of just the same kinds of silicate minerals as other kinds of rock: there is no special ingredient. So how does the spiciness come from the soils? It's not said.

It seems to be taken as somehow self-evident that in the vineyard it is the geology that is primarily responsible for wine taste, and other possible factors are simply ignored. After all, the soil is right there, palpable and familiar. But there's a whole range of other factors involved, which are known to influence grape ripening and hence the flavour compounds in wine – but they are *invisible*. And they can vary at a surprisingly local scale.

Fine variations in bacteria and fungi have now been documented in both the air and the soils of vineyards. For example, the fungi in the West Auckland, Hawke's Bay and Marlborough regions of New Zealand differ from one another, and are quite distinct from those in Central Otago. Fungal communities associated with Uva di Troia grapes in the San Severo Rosso, Rosso Barletta, Cacc'e Mmitte and Tavoliere delle Puglie DOC areas of northern Puglia, Italy, are all distinctive. In California, bacteria on Merlot grapes have been shown to vary between individual vineyard blocks.

Temperature measurements of the soil surface in a Pfalz vineyard at midday varied by five degrees over just a few metres. Closely spaced sensors in, for example, vineyards in Pomerol, New Zealand's Marlborough region, the Columbia Basin of Washington and Oregon, and the Urlaţi, Huşi and Bucium regions of Romania have recorded daily and seasonal fluctuations in such things as air and leaf temperatures, sunlight intensity, humidity and air-flow at a surprisingly fine scale.

A slight change in the orientation of the vine rows, a slight declivity in the ground, a low wall or a few bushes: they all can disturb things. My own house sits on an even-looking hillslope but on a cold winter's morning some parts lose their frost covering way before others. Why? It's by no means obvious. Such

localized variations are in line with 'Segal's Law': a man with a watch knows the time exactly but a man with two watches is never sure. That is, a grower with one weather station may *think* he knows his vineyard's climate but a fine network of sensors may reveal a different story.

As a final example, a geological fault runs right through the south-facing slope of the Abacela Winery in the Umpqua Valley, Oregon, giving marked variations in soil types over small areas, with corresponding changes in the wines. Such wine differences would normally be taken to reflect the geological variations, but in this case, the scientific bent of the owners prompted them to check out the mesoclimate. For five years the owners collected meteorological data from 23 different spots within the vineyard, every 15 minutes. This revealed surprising variations in such things as intensity of solar radiation, and temperatures during the ripening period differing by nearly 5°C – all within this single vineyard. On their concluding list of factors that influenced the ripening of the grapes, the authors did not rate the soil differences highly.

ABOVE: The soils and vineyard walls of Somló, in Hungary: here, the dark rocks soak up heat from the sun's rays during the day and re-radiate it at night, promoting warmth for the grapes and helping the ripening.

A leap too far?

In view of all this, is it enough just to make assertions about geology without offering a basis? Saying, for instance, that an Austrian Riesling wine has 'complexity because of the slatey paragneiss, amphibolite and mica soils' may sound impressive, but surely some justification is needed? Many a winery has samples of its vineyard soils on display, some with enthusiastic claims about the unique contributions they make to the wines, though I've never seen anything explaining how this works.

Of course, relating a wine to its geology has immeasurable marketing value, and bestows the very fashionable 'sense of place'. And, as James Busby pointed out around 200 years ago (*see* Chapter 14), vineyard geology is one of the few things that cannot easily be replicated elsewhere.

Our understanding of how plants grow is already pretty impressive, as witnessed by the success of hydroponics and an amateur's sterile growbags (Chapter 2). And science continues to fine-tune things, helping improve all-round standards of wine production and adding to our comprehension. It continues to reveal to us the sheer complexity and dazzling intricacy of the natural processes involved in making the solid earth yield flourishing vines.

And yet, being science and therefore never beyond question, it is possible we are missing something, overlooking some factor as yet unrecognized that might underpin all the anecdotes on geology affecting taste. I don't know in which direction such a thing might lie but it's possible that the vineyard geology is doing something that at present we are simply oblivious of. Perhaps a reader of these chapters will point it out! That would be tremendously exciting, and then at last I could be jumping for joy.

CLOCKWISE FROM OPPOSITE TOP LEFT: Angular limestone pebbles, Vau de Vay, Chablis, France. Basalt pebbles, Rocks District, Milton-Freewater, Oregon, USA. Terra rossa at sunset, Blok vineyard, Coonawarra, Australia.

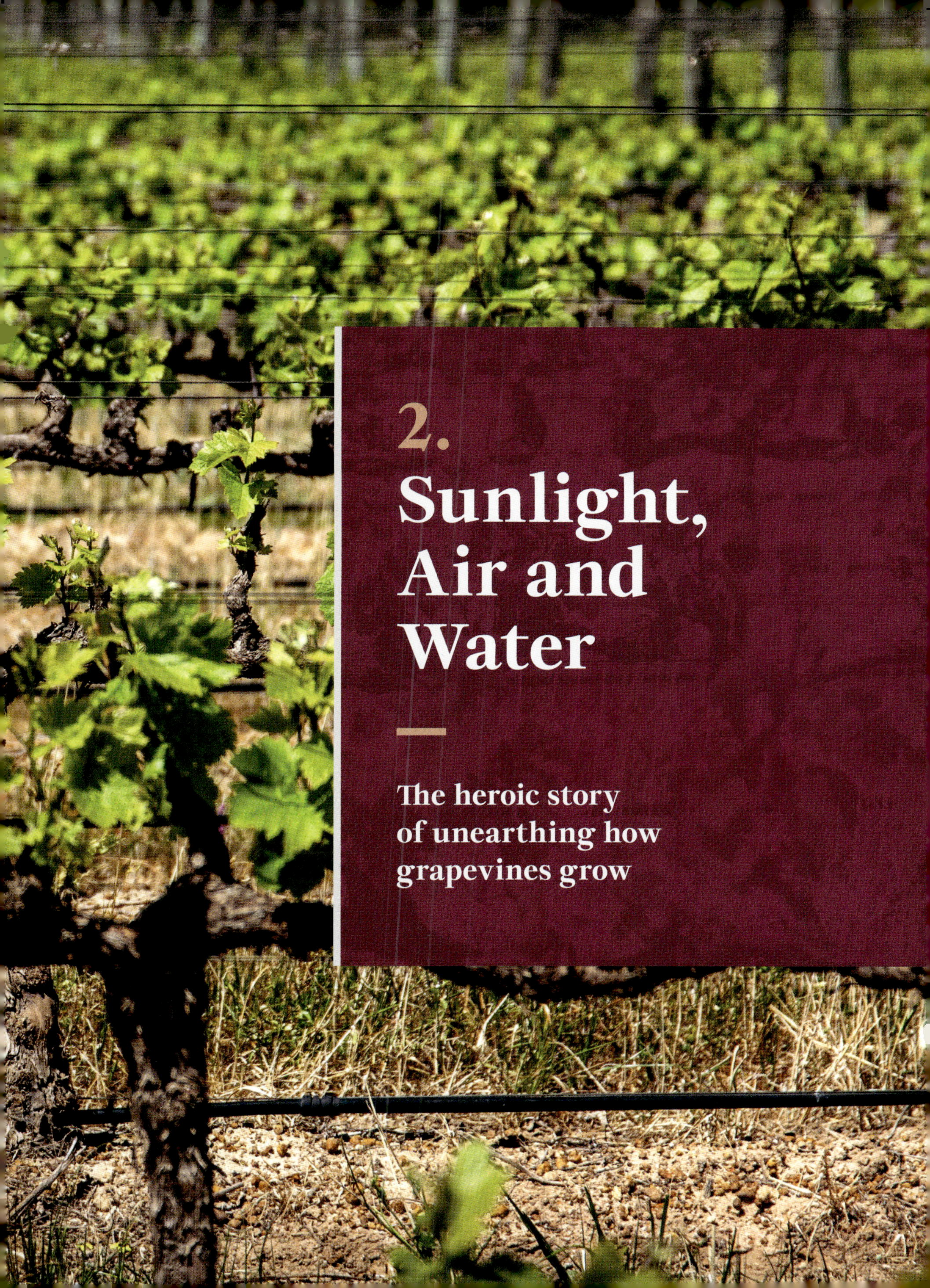

2. Sunlight, Air and Water

The heroic story of unearthing how grapevines grow

Sunlight, Air and Water

Wine is made of soil. It has to be: animals grow by drinking water and eating food, so grapevines, manifestly getting water from the soil, must be getting their food along with it. That is, vines grow by eating the soil. What alternative is there? Consequently, therefore, grapes are made of soil, as is their fermented juice – wine. And that's why wine tastes of the soil and the soil–wine link is so absolute.

It looks like a cast-iron logic and understandably it ruled our thinking for centuries and more, until we discovered there were fatal flaws. We gradually learned of the centrality of gases in the air for vine growth and the criticality of sunlight in photosynthesis. It turned out that grapevines didn't eat soil, after all. Here I outline the more than 2,000-year story of how this gradual realization came about and recall some of those who helped change our thinking. Heroes of science they are, though some, as we shall see, had all-too-human fallibilities.

Soil and wine – gifts from God

The belief of a special relationship between plants and the soil goes back to classical times and way before. Aristotle, for instance, deemed that all matter was composed of earth, air, fire or water, and in the case of plants, of earth. 'Earth is the principal element of which plants are composed,' he wrote around 350 BCE, and went on to argue that this was why they lack feet and the sensory organs of animals: they have no need to move around to seek food. Moreover, Aristotle believed, plants lack digestive organs because their earth diet is 'already digested and prepared' in an earthly stomach, a notion exactly in line with the idea of a 'Mother Earth', an even more ancient and seductive concept and one that for some still enchants today.

In medieval times, not only did it still seem plain that vines grew by consuming matter from the soil, but it was important that the process appeared perfect: differing from animal growth it produced no nasty waste matter. And in those highly religious times such perfection could only mean that plant growth was directly overseen by God. Furthermore, as for how the juice of the fruit of the vine transformed into the life-uplifting liquid we call wine, well, that was nothing short of a miracle, the result of further divine willing. Thus, wine was seen as *goddes god*: Middle English for God's gift to us, from the soil.

The creed was essentially a consolidation of a long-standing religious reverence for the soil and grapevines (apparently the Bible mentions soil 112 times, grapevines 49 times and vineyards 72 times). It was embraced and nurtured by the newly powerful Western Church and the doctrine handed down from generation to generation. And nowhere more so than in 12th-century Burgundy as monks eagerly planted new vineyards, finding the best sites by painstaking trial and error and by careful observation.

The story is frequently and fondly repeated of Burgundian monks tasting the soil to help decide where to plant their vines, but it's probably apocryphal, albeit wholly in line with the beliefs of the time. (As far as I can see, such an endeavour isn't mentioned at all in the thorough and detailed monastic records of the monks' activities; the earliest source I can find for the story is Hugh Johnson's *The Story of Wine*, in 1989, mentioning that 'a modern Burgundian winegrower has said she believes the Cistercians actually tasted the soil').

Nobody expects the Spanish Inquisition

Despite all this dogma, there must have been some inquisitive minds wondering about this soil–plant union, though they would only dare do it in private. To question the teachings of the Church in those days was perilous: it was heresy. But in about 1450 Nicholas of Cusa did just that, though cleverly resorting to that device used by a number of questioning thinkers – Galileo, for instance – in which the radical thoughts appeared in an imaginary dialogue between two fictitious people, the sceptic being cast as some sort of fool.

Cusa, posing as the 'idiot', claimed that he would 'finde the earth but very little diminished' if he weighed it during plant growth, in other words, plants didn't grow by consuming soil matter. Cusa didn't specify any particular plant, though it's fun to suppose that he may have thought of grapevines: he was a native of the Bernkastel suburb of Kues – the origin of his Latinized name – from where he will have looked straight across the Mosel to the celebrated Doctor vineyard, probably one of Europe's most renowned sites even back then.

Nearly two centuries went by before someone openly dared do such an experiment – and he paid the price. Jan Baptist van Helmont (1579–1644) lived in what is now the Dansaert district of Brussels, near the Place du Nouveau Marché aux Grains (Speelplein Nieuwe Graanmarkt). The square today is dominated by parked cars and a kids' basketball court but at one end is a half-forgotten statue to the great man, a free thinker in the early days of modern science.

In around 1629, a time when Brussels was under Spanish rule, van Helmont carried out a simple but historic experiment. He began by weighing a bucket of dry soil and then in it planting a willow sapling. Then he watched patiently as the tree grew substantially in size, out in the rain and the sunshine. After five years he carefully removed the tree from the bucket, shook the soil from the roots and dried it, then returned the soil to the bucket and weighed it all again. He found that the weight was more or less the same as it was in the first place: the plant *had not grown by eating soil.*

So it was that in March 1634, agents of the Spanish Inquisition called at van Helmont's house and took him away for questioning. The interrogation led to his being put under house arrest, charged with 'heresy, impudent arrogance and violating God's laws' by studying the growth of plants and other phenomena. His experiment and its results were suppressed, and they didn't see the light of day until after his death, when his son made copies available. Some found their way to London. There, in a curious twist of fate – divine revenge? – those copies got caught up in the 1666 Great Fire and only one manuscript page survived.

ABOVE: Free thinking about the growth and nutrition of plants could get you into trouble in the 17th century: the statue of Jan van Helmont, Dansaert, Brussels, Belgium.

Before that catastrophe, Robert Boyle – he of the famous scientific law[1] – had carried out similar experiments and got the same general results, on a whole variety of plants. That didn't include grapevines, but his conclusion seemed inescapable. However it is that plants – all plants – grow, it isn't by eating the soil.

So gases are involved

It took more than a century for another significant advance to come, when several individuals began to realize that somehow gases in the air must be involved in plant growth. At that time, the whole concept of gases and the nature of the air around us was new and intriguing. (Oddly enough, the word 'gas', from the Greek word for chaos, was used first used by none other than van Helmont.)

A major figure in these gaseous studies was the Yorkshireman Joseph Priestley, who, living next to a brewery in Leeds, became intrigued by the bubbles in beer. (That led to his inventing carbonated water, which he found 'peculiarly satisfying' to the taste. He even arranged for James Cook to take some of this new-fangled sparkling water on his second great voyage to the Pacific, though Priestley chose not to exploit his invention commercially. That was taken up by a Mr Schweppe…)

Among many other contributions, Priestley went on to isolate the gas we now know as oxygen. Famously, he burned a candle in a closed jar until the air inside could no longer support combustion and reasoned that the gas required for the candle to burn, oxygen, had been used up. What is important in our story, however, is that he also did the experiment with a living plant (mint) in the jar and found that the candle burned for longer. Therefore, he argued, the plant must be giving out oxygen. So plant growth somehow involves gas. Who would have thought it?

And so is sunlight

The next breakthrough in the story happened by pure serendipity. And once again it involved an individual with multifarious interests. Jan Ingenhousz was born in Breda, the Netherlands, in 1730, but his fascination with the works of

1 Robert Boyle was a 17th-century pioneer of modern chemistry, best known for showing that the volume of a gas decreases with increasing pressure and vice versa – the famous Boyle's law. (Or, as Flanders and Swan whimsically rephrased it for politicians, the greater the external pressure the greater the volume of hot air.)

Joseph Priestley (and with the newly discovered technique of inoculating people against the scourge of smallpox) took him to England, where he settled for the rest of his life.

Much of his time was spent in London but in 1779 he 'disengaged from the noise of the metropolis and retired to a small villa, out of the way of being interrupted in his contemplation of nature'. It was at Bowood House in Wiltshire, through no coincidence the same place where his much venerated luminary Joseph Priestley had lived some years before. By now Ingenhousz was intrigued by how plants grew and one day he picked living leaves from the garden and took them to his study – the self-same room where Priestley had carried out his famous candle-in-a jar experiments.

He put the leaves in glass jars on the table – *under water.* To his surprise, he saw little bubbles around the leaves in one of the jars: surprise not at the bubbles for he knew well of the involvement of gases, but that it was happening in only *one* of the jars. Why not the other ones? Then he realized that the bubbling leaves were catching a ray of sunshine coming through the study window. Out of the sunshine the bubbles stopped, and even taking the jar over to the warmth of the fire in the corner of the room didn't trigger them.

It was, he reasoned, the sun that was causing the bubbles to form, not through its warmth but the actual sun *light.* Growth of the leaves, he wrote, 'is not owing to the warmth of the sun but chiefly, if not wholly, to the light'. We had, of course, always known that plants couldn't grow in the dark, but Ingenhousz had discovered that for plant growth, sunlight, what today we would call ultra-violet light, was an intrinsic driving force.

Progress indeed, but the questions kept coming: what was it about sunlight? How did leaves retain their size and shape if they were losing oxygen? The latter question was answered by several workers, such as the Swiss Jean Senebier in 1782, by showing experimentally that the oxygen loss was being countered by plants *taking in* a different gas. Senebier referred to it as 'fixed air'; van Helmont had called it '*gas sylvestris*'; Ingenhousz had written of 'bad air' associated with leaves and fruit. These were all names for the same gas and today we call it carbon dioxide.

OPPOSITE: Benefiting from autumn sunlight: the author's own Seyval Blanc grapes growing in Wales.

Nutrients from the soil

So plant growth involved their taking in carbon dioxide from the air. Then the Swiss scholar Nicolas Theodore de Saussure in 1804 showed that there must be other things involved as well – he talked of potash, lime and magnesia (most of the actual chemical elements were still unknown at that time). By weighing plants in controlled experiments, he demonstrated that these were not 'accidental ingredients': although the amounts needed were tiny, they were *essential.* Moreover, he showed that roots don't allow 'any and every nutrient' to enter the plant; they have a 'special action' that allows different nutrients to be absorbed to different extents. Perceptive stuff, often overlooked today.

In 1834, the Parisian Jean Baptiste Boussingault began further experimental measurements on plant growth, now not in a laboratory but out in nature, for the first time ever *in the field* (literally!). His farm was in Alsace, not in the celebrated wine growing district but up at Pechelbronn ('pitch spring'), where for centuries the locals had exploited the tar sands. Boussingault himself had once worked asphalt there and in 1833 he returned to Alsace to marry. Through this he became joint owner of the Pechelbronn farm where he established whole series of field experiments, with control plots and replications, weighing and analysing soils, manures and crop yields. They all established that indeed these nutrients were essential to growth, that they came from the soil and, importantly, that they included nitrogen.

And in the vineyard?

So what did all these breakthroughs in understanding mean in practice for a winegrower? Well, at first not much, apart from at a few illustrious and well-connected estates, say, in Bordeaux. Most viticulture was a peasant activity with workers toiling simply to earn enough wine to make a living, in just the same way as their forefathers. And the vines did what vines do – scramble and twine across the ground in the absence of trees to climb up. The result was a jumbled mass spreading by 'layering', just as brambles and briars do, giving poor access for sunlight. Down in the soil was a tangled mass of roots, badly inhibiting drainage.

One person more than anyone attempted to improve all this on a scientific basis: Jean-Antoine Chaptal. Besides a host of general contributions such as introducing the word 'nitrogen' (in French *nitrogène*) and the metric system of measurement, he was the first major science-based writer in France to tackle

improving wine. He implored the wine growers to seek quality rather than quantity and although most of his work concerned winemaking, in the vineyard he argued for more order, for spaced vines with vertical supports, carefully trained and pruned.

The soil must have good drainage, he asserted. Clay was to be avoided if possible, but gravel, limestone, granite and volcanic soils were all suitable bedrocks on which to plant vineyards. The actual nature didn't matter, he argued,

ABOVE: Grapevines carefully spaced and trained with a view to yielding good-quality fruit, as advocated by Jean-Antoine Chaptal, at Villány, Hungary.

as long as the soils were light, porous and friable, like those of Bordeaux and Burgundy. (But *see* Chapter 12 for the contrary views of James Busby.)

However, the thing that really put science into the minds of artisan winegrowers was the arrival later in the century of phylloxera. Suddenly their very livelihoods became threatened, and they turned to science for help. Until then, university and state funding for viticultural research had all been rather perfunctory, but suddenly it was needed. It's telling that, for example, when in 1868 wine growers in the Languedoc discovered an insect devouring the roots of their vines they went to the local university (Montpellier) for advice – to the department of pharmacy!

The nitrogen fix

Both de Saussure and Boussingault had demonstrated that nitrogen was an essential ingredient of plants and that it derived from the soil, not the air. This, however, was fiercely disputed. The air is, of course, around 78 percent nitrogen, and various workers attempted to argue that this must be where it came from.

A somewhat forgotten individual, Philipp Carl Sprengel, born near Hannover, Germany, in 1787, had also concluded that nitrogen came from the soil and had even extended the knowledge into the practice of adding it (in the form of ammonia, NH_3, compounds), together with other nutrients in the form of ground-up marl, gypsum, limestone, etc, to the soil if it seemed needed. In other words, he had invented fertilizer.

It was, however, a fertilizer protagonist much more illustrious (partly thanks to his own machinations) who put the practice onto a commercial footing while vehemently disputing the soil source of nitrogen. This was Baron Dr Justus von Liebig. He was born in 1803 at Darmstadt, Germany, where his father owned a drugstore that sold things such as paint and varnishes made on the premises. It gave young Liebig his first experiences with chemistry. Then, among many later endeavours, he developed precise methods of analysis, some involving bizarre equipment such as the connected glass bulbs he used to analyse hippuric acid in horses' urine. Older readers may recall 'Liebig condensers' from chemistry classes.

Without doubt, Liebig's contributions were seminal; he's been called 'the founder of organic chemistry', the 'father of the fertilizer industry', and so on, but he was prone to self-aggrandisement and speculation rather than careful measurement. He enthusiastically promoted the work of others – without mentioning their names; one critic wrote that 'he drew his conclusions not in his laboratory, but at his writing desk'. Thus he was, as another commentator wrote: 'More a promulgator and defender of truths that had already been announced than a discoverer of new knowledge.'

Liebig's self-assurance is well illustrated by his very public and vituperative feud with the Englishmen Lawes and Gilbert. Sir John Bennet Lawes was a well-to-do landowner intrigued by agricultural methods such that in 1843 he turned over some of his estate to scientific research. It was the founding of England's now celebrated Rothamstead Research Institute. Lawes saw the need for a proper chemist to join him in his venture and so appointed the young Joseph Henry Gilbert. Ironically, as it turned out, Gilbert had obtained his PhD in chemistry from the University of Geissen – home of Liebig's far-famed laboratory.

Lawes and Gilbert systematically investigated separate test plots over some years and they always found that paucity of nitrogen was the critical cause of soil depletion. Conversely, its availability in the soil rapidly augmented plant growth. The air had nothing to do with it. Lawes and Gilbert thought their results 'so decisive' that it was 'hardly possible to have two opinions on the subject'. But Liebig thought otherwise and was incensed. He declared the work 'most destitute of probability' and 'entirely devoid of value' (though one German critic pointed out that Liebig himself 'had not raised one grain of wheat').

A visitor to Rothamstead noted that the sheer rigour of the work there made it our 'principal source of trustworthy scientific information on agriculture'. But Liebig would have none of it: 'The experiments of Messrs Lawes and Gilbert are very far indeed from proving the conclusions which they wish to draw; they establish rather the fact that these gentlemen have not the slightest notion of what is meant by argument or proof.'

Despite all the bluster, however, eventually the evidence against Liebig's position became simply overwhelming. Demonstrably, plants got their nitrogen from the soil, not the air. So was the great man humbled? No, he breezily began promoting Lawes and Gilbert's conclusions with no mention of them or his previous utterances. And jubilation for the English two? Lawes and Gilbert graciously offered that 'to the generalizations of Liebig founded nearly a quarter of a century ago, we must attribute much of the stimulus and direction that has been given to chemical inquiries in connexion with agriculture in recent times'. All worthy of the smoothest of politicians.

Complete, balanced nutrition

Julius von Sachs was born in the Silesian city of Breslau (now Wrocław, Poland) in 1832, and apparently was a shy, introverted child. Near his house was a large garden where he enjoyed escaping to observe flowers, collecting, cataloguing, pressing and cultivating them and, especially, learning to draw them in fastidious detail. In fact, when at age 17 von Sachs became orphaned through a combination of strokes and cholera in his family, it was income from his skill at illustrating that sustained his studies through college.

He then moved through a succession of academic positions, ending with being elected Rector of the University of Würzburg, where the local government bestowed an order of nobility upon Sachs, thus permitting the addition of 'von' before his family name. He became obsessed with his work. So much so that, together with losing his wife to a mental institution, heavy smoking, and lack of exercise, he became wholly dependent on medication to survive. But throughout, his output was immense, seminal and diverse; two of his contributions – the roles of nutrients and of light – are particularly relevant to our story.

ABOVE: Air, water, sunlight and minimal nutrients: vines growing in a hydroponic system without recourse to rocks and soil. How much geological input does a plant really need?

First, plant nutrition. Consider. A vine planted in moistened ground-up rock and with access to sunlight and air will survive for a while, but it won't thrive. Rather like a cut flower in a vase of water, it will eventually give up. It needs, as we have seen, nutrients. And von Sachs devised a way of growing a plant with its roots in water, in which he could dissolve one by one the chemical elements known at the time, and so determine which ones were essential for growth and in what amounts. He established that about 17 nutrient elements were needed by most plants (although three of them, selenium, sodium and aluminium, aren't needed by grapevines).

And here's the important thing for us. If we get one of these elements, say iron, grind it up and scatter it on the vine, it will make no difference. Spreading the pulverized iron on the soil won't help either, even getting it down next to the roots. Von Sachs demonstrated that to be usable by a growing plant, *the nutrients have to be dissolved in water*. Apart from, we now know, the functioning in some situations of mycorrhizal fungi in the soil[2], vine roots simply cannot take in solid material.

Moreover, he showed that plants need nothing further beyond these nutrients, given air, water and sunlight. Unlike animals, there's no need for vitamins, proteins, fats and the rest. It's all confirmed today by the success of the hydroponic method of agriculture (supplying plants under UV light and in a sterile substrate with water containing a dissolved cocktail of nutrients), now probably much more commercially important than many people realize. Indeed, many an amateur gardener will enthuse about the superiority of their homegrown produce, tomatoes say, which are more likely than not grown in a sterile growbag – not soil – together with water and formulated nutrients out of a bottle. It works, and it was all made possible by the work of von Sachs.

Incidentally, I should also note that in 1860 Louis Pasteur explained the transformation of grape juice into wine by yeasts fermenting the sugars. Yeasts, just like grapevines, are made very largely of carbon, oxygen and hydrogen, so it's worth reminding ourselves that these three elements comprise more than 99.5 percent of a wine, albeit in a bewildering array of combinations. And we now know that it's these compounds, some of which are strikingly aromatic and flavoursome, that make wine taste like it does.

2 Mycorrhizal fungi produce enzymes that enable them to extract nutrients directly from solid rock. Then, living intimately with a plant's root system, they deliver these nutrients to the roots in exchange for sugars, fats and the like produced by the plant. Soil tillage and herbicides – common in vineyards – destroy the fungi.

Storing the light

Where light is absorbed by certain metals, electrons can be dislodged and form an electric current (as happens in solar panels). Where plants absorb light, they instead produce pigments, the best known of which is the chlorophyll that gives leaves their green colour.

This was all worked out by von Sachs in about 1862, and marks the second of his seminal contributions to our story. He showed that the chlorophyll occurred in special bodies in leaves and stems and, driven by the absorbed light, took in carbon dioxide to make starch. This, together with glucose, enabled the plant to convert sunlight into energy and to store it, and form the starting point for all the chemical reactions that follow, in other words the plant growth.

So finally, things had come together to give a coherent understanding of how plants grow. There were loose ends of detail to tie up – and still are, down at the molecular level – but the central thread of our long story has almost reached its end.

Photosyntax anyone?

Charles Barnes was born in the charming old riverport of Madison, Indiana. Thirty-five years later, in 1893, he found himself in the city of Madison, Wisconsin, addressing a meeting of the Botanical Section of the American Association for the Advancement of Science. His talk was entitled 'On the food of green plants', and in it he summarized our scientific understanding: stored sunlight drives the intake of water and carbon dioxide, in the presence of certain nutrients, from which compounds of carbon, oxygen and hydrogen are derived. Quite a mouthful.

So Barnes argued that a single word was needed to cover all this and he proposed 'photosyntax'. Or, he added, an alternative though less preferable word would be 'photosynthesis'. It is, of course, this alternative that caught on. I suppose that these days most folk have at least heard of photosynthesis and many probably know that it's how science explains the growth of plants. It's been a long journey, but I don't know of anyone today who disputes that plants grow by photosynthesis. After all, the theory is backed by overwhelming evidence – theoretical, experimental and observational. And yet, and yet.

In modern wine writings I read things like 'the vines clutch the earth and impart its essence' and 'I will never be able to eat soil like a plant can, but wine allows me to taste soil and bedrock'. Is this thinking really that different from that

of the Middle Ages? Perhaps the idea that wine is of the soil is still with us? Of course, soil is fundamental to vine growth, especially through its drainage, water-holding properties and nutrient provision, but it strikes me that those millennia of pre-photosynthesis belief are maintaining a powerful legacy. Surely nobody these days thinks that vines grow by eating the soil. Do they?

ABOVE: Madison, Indiana, USA, home of the American botanist Charles Barnes, who became professor of plant physiology at Chicago University and who coined the term 'photosynthesis'.

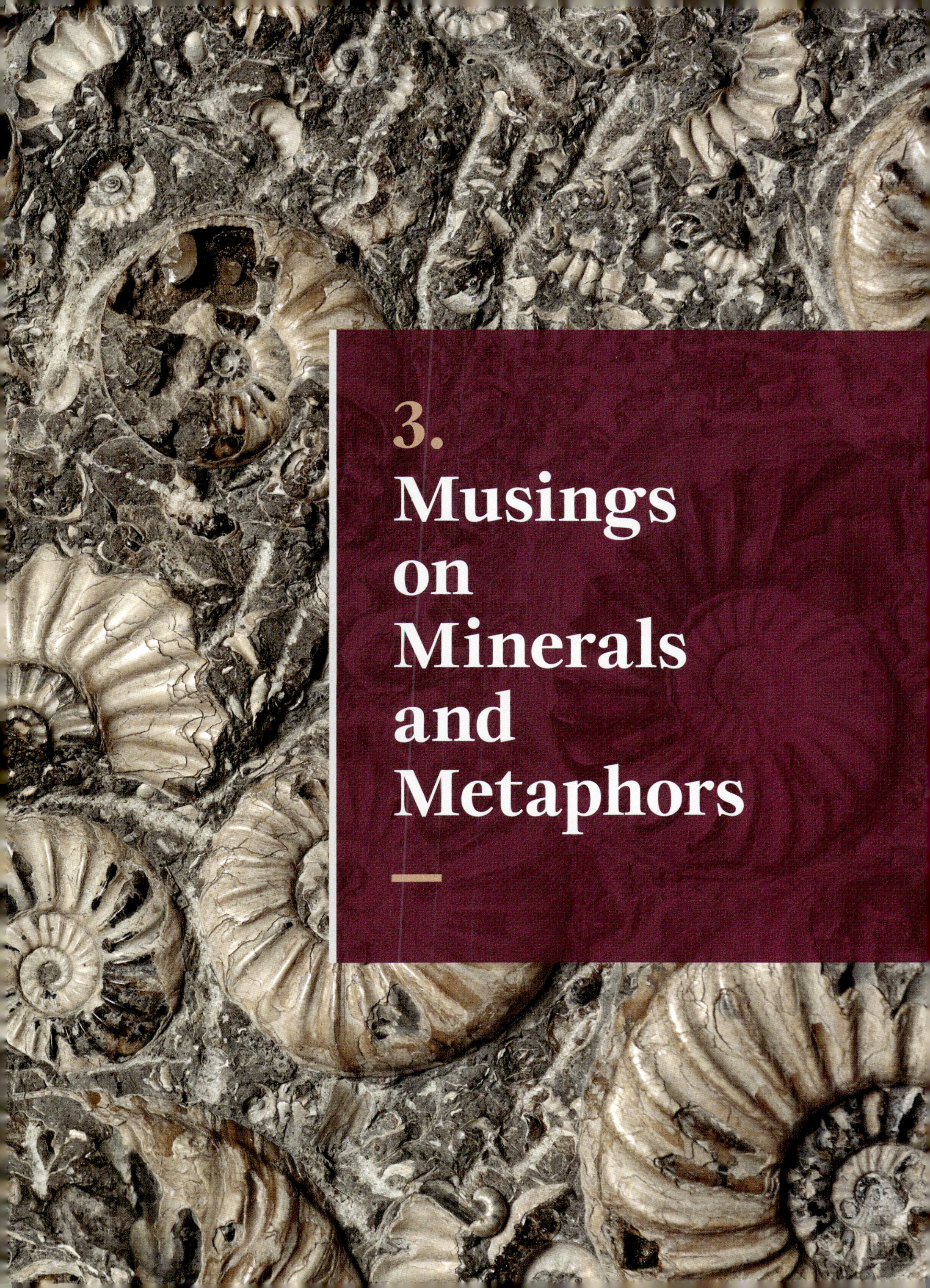

3. Musings on Minerals and Metaphors

Musings on Minerals and Metaphors

It used to be simple. The word 'mineral' was either a noun referring to things found in the ground or an adjective for something to do with the ground. It wasn't an alluring word: for geology graduates, I remember, the minerals industry was a major employer, but I don't think that working with coal, iron ore and the like was ever seen as remotely glamorous.

In the last two or three decades, however, things have changed. Mining is directed at exotic-sounding materials used in fancy technologies, and elsewhere the word has taken on a range of new senses, some of which involve desirability and even charisma. Unfortunately, the new meanings are frequently blurred, and 'mineral' has become a confusing word. And nowhere is this more the case than in the world of wine. Here I take a closer look at this newly muddled world of minerals, beginning not with wine but with two curious parallels.

Mineral waters

Charles Dickens, Florence Nightingale, Lord Tennyson – they all visited the English town of Malvern for the same reason, to sample the supposedly curative waters that flowed from the ground there. And, unusually for the 19th century, such visitors to Malvern could take some of the water away with them when they left – in 1622 it had become the first place in the world to put its water into bottles. (Much later, a Mr Schweppe went to Malvern and turned the idea of bottling its water into a major commercial enterprise.)

What was it about that Malvern water? It all had an air of mystery, in line with the myth and superstition that springs and wells had always attracted. Today, supermarket shelves sag under rows of bottles labelled 'mineral water' and perhaps there's still just a trace of mystery, or confusion at least. For instance, it seems that some folk buy it because it's perceived as healthy and 'pure', unlike tap water, while others think it contains an array of health-giving substances that tap water lacks.

You might think that 'mineral water' implies a water that has something to do with the ground, but that's not necessarily so. (Even as a young boy I was puzzled at why soft drinks like dandelion and burdock were called mineral water.) Today, depending on the jurisdiction, the term can be legally defined as 'a water that contains salts and gases' or 'a water with a dissolved ingredient content of more than 1,000 mg/l', in some cases specifying the dissolved elements. Some definitions still involve the traditional curative beliefs: 'Mineral waters are those with therapeutic properties and acknowledged as such by the Medical Academy of France.' None of these mentions the ground. In other words, for water the word 'mineral' now doesn't seem to mean very much at all. It just adds, well, something.

Mineral cosmetics

Iconic Egyptian queens such as Cleopatra and Nefertiti outlined their eyes with ground-up geological minerals, things like stibnite (antimony sulphide), galena (lead sulphide) and pyrolusite (manganese oxide). Elizabethan ladies (and gentlemen) whitened their faces with pulverized cerussite (lead carbonate); in Victorian times, cheeks rouged with ground-up litharge (lead oxide) and cinnabar (mercury sulphide) tickled fancies. The idea of applying such toxic substances to your skin might seem horrifying today but ground-up geological minerals are present in most modern cosmetics. Examples include thickening ingredients such as sericite mica (magnesium aluminium silicate), whiteners such as ilmenite and rutile (titanium dioxide), quartz (silicon dioxide) for texture, and muscovite mica (potassium aluminium silicate) for shine.

But in an interesting parallel with the wine situation, in recent years this mineral content of make-up has been handled in a cleverer way and with a blurring of meaning. It used to be that the mineral ingredients of cosmetics were left unsaid but now their presence is actively promoted. It seems that mentioning minerals in cosmetics now adds a certain cachet, a selling point.

One tactic is through the minerals being touted as *natural*. No synthetics, the marketing blurbs proclaim, just 'minerals extracted from our precious Earth'. In other words, just as with wine, a perceived link with the land is seen as special – as in 'our cosmetics are infused with the essence of Mother Earth' – and maybe even spiritual: 'Every mineral used in our cosmetics is special and transfers its natural energies to our mind and our beauty.'

And also as with wine, the distinction between geological minerals and nutrient minerals (dissolved single elements) becomes lost. Our skin needs certain mineral nutrients (such as zinc, copper and selenium) which normally are taken in as part of our diet, but modern advertising suggests that these and a whole range of supposed skin nutrients – casually referred to as minerals – can be directly absorbed from geological minerals in make-up. My point here is not the plausibility or otherwise of all this but to illustrate the newly exalted and blurred uses of the word mineral.

That's exactly what has happened in the world of wine. In my wine books from the 1960s and '70s, by authors such as Michael Broadbent, Alexis Lichine and Cyril Ray, I can't find the word mineral at all. In marked contrast, today the word is almost omnipresent in writings about vineyard rocks and soils, vine nutrition, grape juice, wine composition and the very taste of wine. Yet its meaning is often vague.

'Mineral-rich' soils?

To emphasize the distinction between geological and nutrient minerals (*see* Chapter 1): a breakfast cereal, for example, might be 'enhanced with minerals'. This doesn't mean there's pyrite or feldspar mixed in with the cornflakes; the analysis will be telling us how much iron, sodium or zinc – single elements – the packet contains. If we go to a health food shop and ask for minerals, we will be shown jars filled not with such things as quartz and mica, but tablets containing particular elements – magnesium, calcium and the like.

The trouble is that in wine writings the two meanings are often confused. And the most common instance of this, which I see all the time in wine web pages, magazines and so on, is in mentions of vineyard soils.

Wine commentators often state that a vineyard soil is 'mineral-rich', usually implying that this leads to desirable qualities in the resulting wine. I see vineyards lauded as being 'rich in minerals', 'laced with minerals', and superior because they're 'blessed with rich minerals'. These are seductive phrases, but what do they mean? All rocks and soils are made of (geological) minerals, not some more than others. There may be a greater *diversity* of the minerals, such as in volcanic

OPPOSITE: Gravel soils at Château Beychevelle, Saint-Julien, Médoc, France. The stones provide excellent drainage for the vines and are claimed to give 'distinctive minerality to the wines'.

soils relative to limestone, but without shutting down the soil's pores – not a good thing – one soil can't have *more* minerals than another.

So maybe it means rich in nutrient minerals? But that's the same as saying fertile, and it's pretty much axiomatic in viticulture that rich soils are to be avoided – some of the most revered wines come from vineyard soils that are astonishingly barren looking. Fertility is music to the ears for a farmer growing 'greedy' crops such as sweetcorn or tomatoes, but with vines it leads to high vigour, lower grape quality and consequently poor wine. Thus the word mineral in these phrases has no real meaning even though it sounds good. Rather like with 'mineral water', it somehow adds desirability; it's as though this one word lends a new, transcendental meaning.

Nutrient minerals are, of course, absorbed by the vine, but what this means for wine is complicated. For instance, the much-loved notion that mineral nutrients are taken up by the vine roots and eventually find their way into the finished wine for us to taste, is actually a very convoluted affair.

For starters, the roots don't simply soak up like blotting paper whatever happens to be available in the soil. Rather, they have an armory of intricate and sophisticated devices aimed at screening out undesirable elements and selecting each of the 14 or so essential nutrients in the optimal amounts needed for healthy growth (Chapter 2 recalls the discovery). The mechanisms, though, are far from infallible. Some of the dissolved nutrients will find routes that bypass the screens and be passively absorbed. Some nutrients will interfere with each other; some will conspire, as it were, to fool the vine's systems.

And that's why growers assess nutrition by analysing not the geology but the vines themselves, to see what they are actually taking up, correcting levels as necessary. It's why a good grower walks his vineyard: the key to monitoring its health lies in watching the vines, not the soil.

In practice most of the nutrients come from the soil's humus (*see* Chapter 15) though the vine does not care, so to speak, whether a particular nutrient mineral originated in humus, as this or that geological mineral, or in a bag of fertilizer. Magnesium is magnesium irrespective of its source and does the same jobs. Moreover, an absorbed nutrient is apportioned by the vine around its component parts – different proportions to the leaf tissues, to the stems, to the grape skins, seeds and flesh, etc. The upshot of all this is that generally in practice the (nutrient) mineral content in the grape juice depends little on the vineyard geology.

Tasting minerals in wine?

The proportion of nutrients in the grape juice changes as winemaking proceeds. Fermentation, filtering and fining all remove some elements while adding others. Changes can arise from contamination in the winery, such as from pipes and fermentation vessels, and during maturation. And that's why the proportions of mineral nutrients in a finished wine have only a complex, indirect and distant relationship with the geological minerals in the vineyard. (Incidentally, attempts to 'fingerprint' the origin of a wine with inorganic components have used traces of things such as rare-earth elements and isotopes that, unlike the nutrients, are not subject to selectivity by the vine.)

Some commentators describe wines as 'mineral crammed', 'mineral laden', 'brimming with mineral' and the like, which might sound literal but the actual nutrient mineral concentrations in wine are minuscule. In fact, higher amounts can present problems for the winemaker such as stuck fermentations and hazes.

Nevertheless, nutrient minerals, of whatever origin, are there in wine and so it's interesting to ponder if they can be tasted. Probably not, because the detection thresholds for inorganic elements in drinking water are in most cases higher than the concentrations normally found in wine. Moreover, tasters report that as the values increase above that threshold, the taste of the water becomes more and more disagreeable. And because in water there are few competing flavour compounds, the thresholds in the presence of the hosts of aromatic organic compounds that define wine must be vastly higher in wine (and hence, presumably, even more distasteful).

The presence of these minerals in wine can collectively influence our taste perceptions but that's not the same as tasting an individual mineral, let alone identifying it. I once thought of having some undergraduate students taste a series of wines to which I was adding increasing amounts of various inorganic elements, to see how much was needed to be tasteable. I didn't. Potential litigation is a concern these days in university activities, and it's quite likely that the amounts needed would verge on the toxic…

So where does this put all those casual references to a wine tasting of minerals, and the like? It can't be literal. The idea has to be in our minds, it has to be a mental association, perhaps subliminal. If we describe a wine as tasting mineral, we have to be in the realm of metaphor.

Metaphor: breathing life into language

We're all aware of the difficulties of putting our taste sensations into words and that we often have to resort to using metaphors. And without even thinking about it we know that just as a 'heart of gold' or 'a couch potato' aren't literal, neither are, for instance, plummy or leathery tasting wines. The wines don't actually involve plums or leather, obviously. Except, it seems, where tasters use 'earthy' terms such as mineral. It's then so tempting to cling on to that implied connection with the soil.

Much has been written – ever since Aristotle – about the nature of metaphor but it occurs to me that we use wine-taste metaphors in three different ways. In the first we make a comparison between the wine and something else where, although we may not know it, science has found that there happen to be flavour compounds in common. I don't imagine that most tasters sensing, say, blackcurrant notes in a Cabernet Sauvignon are aware that exactly the same compounds producing that taste in blackcurrants can be in wine. It's still a metaphor, but it has a tangible basis; there is a demonstrable commonality. The 'black pepper' taste of Shiraz and the compound rotundone found in black pepper is another example.

Second are the more generally used metaphors where, as far as we know at present, there are no such direct links but useful recollections of other flavour experiences. Tasting a Barolo wine may remind us of a forest floor, for example, a Marlborough Sauvignon Blanc of grass clippings, that kind of thing.

Earthy-sounding examples of this include referring to the smell of warm or wet stones. We are all familiar with the odour of a shower of rain falling after a dry spell and it's a useful flavour comparator – but it has nothing to do with geology. It makes no difference whether the rain is falling on this or that rock, concrete, tarmac, or whatever, because the smell is due to organic compounds.

Any stone-like surface exposed to the air soon becomes veneered with all manner of bacteria, algae, moulds, lipids and the like, which are all around us. Some settle in the tiny depressions of the porous surface and quite literally – it has been filmed – a falling raindrop can displace the compounds into the air as an aerosol, producing the well-known odour. It has been labelled petrichor.

Similarly, tilled earth and damp cellars produce familiar smells – known as geosmin – also of organic substances that emanate from certain bacteria and moulds. (Some of these have astonishingly low sensory thresholds, down to a few parts per trillion.) We may smell them when soil is being tilled, and we recall this when describing a taste sensation in wine. They are metaphors.

The third way is a variation that's intriguing to me because it's where most of the 'earthy' metaphors fall. Here, we make the same kinds of comparisons, but *with something that itself has no flavour* – no taste or smell – or with something that we have no experience of. (One example of this is flint, which I examine in Chapter 13.) How can this work? I can only think that some constructive imagination must be involved – what the taste of a rock or mineral would be like if it had one. Here are some illustrations of what I mean by this third group.

Slate

Above my head as I write this here in my Welsh farmhouse, is a roof covered with slate. It was put there nearly 200 years ago and is still almost as good as new – slate is such a good roofing material. Why so? One reason is its characteristic property of splitting cleanly into thin but robust sheets. Slate is a rock, a collection of several different minerals bonded together, and here they are arranged in such a way as to allow this useful splitting (*see* Chapter 7). The second reason is its durability and, in particular, its being waterproof. It doesn't react with water; it doesn't dissolve. That's a pretty important property given the damp Welsh climate but it's also a critical matter for us here talking about wine.

It means that no matter how much slate there is in a vineyard soil it simply cannot get into the vine. Roots can only absorb dissolved things (apart sometimes from elements in active mycorrhizal fungi) and slate is wholly insoluble. So the idea that vines take up slate from slatey vineyard soil, and then transmit it through to the grape juice to end up in the wine patently cannot be true.

This insolubility also means that slate, like other rocks, has no taste. It's easily tested if you have access to a rock saw and some different kinds of rock. Smooth, freshly sawn surfaces will give a cool, tactile sensation on your tongue but they will have no aroma or taste. If you lick and smell them blindfolded you won't be able to tell the rocks apart.

In slate-dominated areas such as Germany's Mosel Valley it's easy to believe that all that slate must influence the wine, and almost certainly it does – indirectly. Things like the free drainage, light reflectance, heat retention, nutrient availability may all play a part in the grape formation and possibly a wine's character, but I am talking here about the flavour of wine in the glass, and the notion that the slate itself can be detected in the wine.

Our tastebuds can deal only with liquids. Certain solids may be sufficiently soluble for our saliva to almost instantaneously begin dissolving them and transporting them to our taste receptors – sugar and salt are familiar examples. But slate is wholly insoluble. Even if somehow slate in a vineyard soil did magic itself into a wine glass, then we would see it, and the idea of little specks floating in the wine is manifestly absurd.

So, our reporting a taste perception as 'slate' must therefore involve some mental inventiveness. We may think of slate as a cold material and we may associate a sharpness with its splitting into sheets, such that a slatey taste is almost always described from cool-climate, usually white, wines. And if it's a Mosel wine and we know of the importance of slate in the region it becomes easy to conjure up the slate descriptor. Could we be inventing a sense of what slate *ought* to taste like?

ABOVE: Slates (from Penrhyn Quarry in North Wales) on the roof of the author's house. Welsh slate was used by the Romans and in Victorian times was said to be 'roofing the world'.

A mineral medley

These days it's fashionable in wine-tasting notes to specify an actual mineral. Let's take as an example the mineral talc, as in: 'The wine shows strong *talc* notes' or it has 'a distinct *talc* minerality'. But just as with the minerals that come together to form the rock slate, almost all individual minerals, including talc, are insoluble and hence have no taste. (The only significant exception is salt – sodium chloride – the mineral halite.) So, again, using these mineral names as descriptors must involve mental creativity, but it can get quite confusing. For one thing, the abstract picture in one taster's mind might not correspond with that of another.

Talc is a mineral found in some highly magnesium-rich rocks. Its characteristic feature is a unique softness: it's the softest mineral known. Thus, mentioning talc as a mouthfeel is understandable (although wine having 'a talc-like friction giving the wine backbone' is puzzling), but what about the more common use of talc in tasting notes – as an *aroma*?

Just as things have to dissolve to have taste, so they have to vaporize to have a smell. Aromatic solids or liquids turn to vapour easily but geological minerals like talc simply do not. They have no smell (odour or aroma). However, there's a complication with talc. Its softness is, of course, utilized to produce the well-known toiletry talcum powder, and this usually has had perfume of one sort or another added to it. So as I see it, phrases like 'the wine opens with aromas of talc' or it 'has light scents of talc' are alluding to the toiletry and its perfume, and not to the mineral at all.

It's a rather parallel situation with graphite, the crystalline form of carbon found in some rocks. It also is soft and with a greasy feel such that it's used as a mechanical lubricant. So we might expect that mentions of graphite in tasting notes would be referring to mouthfeel, but they usually allude to taste or aroma, as in 'the soils provide a graphite taste'; 'gravelly soils provide a palette of graphite notes', even though graphite doesn't dissolve or vaporize.

The complication here, of course, is that graphite is also used for pencil 'lead'. The soft smudgy graphite is enclosed in wood and this usually happens to be strongly aromatic. Cedarwood is traditional, and its distinctive sweet notes can often be apparent in wines such as oak-aged Cabernets. But surely this should be noted as a cedar aroma, or perhaps pencil shavings, rather than graphite? It seems to me that a tasting note would be more precise and therefore more helpful in painting a picture of a wine if it said perfume of talcum powder or aromas of cedarwood rather than talc or graphite. But then that charismatic allusion to a mineral, and beyond that the vineyard soil, would be lost…

For some commentators, chalk in a wine is an odour ('the chalk aroma of a Ribolla Gialla'), for others a taste ('the chalk-like taste of fine Chablis'). Both, again, have to be creative metaphors seeing as chalk is a rock, wholly tasteless and odourless. Geologically, it's a particular kind of limestone – and hence made of the mineral calcite, calcium carbonate – that's dominated by the fossilized remains of a certain, very tiny, marine organism.

Most often, talking of chalk in a wine is referring to a grainy texture, a mouthfeel, perhaps in line with a perception of chalk as a dusty powder. Chalk, however, is not necessarily dusty. Some of the best-known natural occurrences of chalk are distinctly hard – think of England's White Cliffs of Dover or Beachy Head. For that matter, neither is it necessarily calcium carbonate.

I'm guessing that the 'dusty' image comes, at least for those of a certain age, from old classrooms and their chalkboards, but blackboard chalk is calcium sulphate, the mineral gypsum. The famous chalk dust of disputed line-calls in tennis is titanium dioxide, the mineral rutile. In other words, a chalkiness in wine is an imagined mental association, with no obvious relationship with vineyard geology.

Now, Chablis and iodine. Most wine enthusiasts know that the bedrocks at Chablis formed on an ancient sea floor and contain striking fossilized seashells. It all puts you in mind of the sea! Consequently, things maritime often appear in tasting notes on Chablis, such as an iodine note. For example, 'Chablis – you can almost chew the iodine'; '…plenty of Chablis character, in particular iodine'; '… that exclusively Chablis note of iodine'. And, naturally, it's usually related to the soils: 'fossilized oyster shells give Chablis its unique iodine flavour' and the like. So is this metaphorical, or perhaps here something more literal?

One complication is that few of us have actually smelled, let alone tasted, iodine. It is an element that can be made to vaporize so it does have a smell – which the books say is repellent and toxic. What most of us think of as iodine is actually a medicinal tincture, with a strong smell of some organic solvent such as alcohol. It isn't the aroma of iodine.

Iodine is a well characterized chemical element that can be measured, so some years back a student (Li-Ming Tan) and I analysed iodine in soil and vine-leaf samples from Grand Cru sites together with finished Chablis wines. Then, for comparison, the same down the road with soils and Chardonnay vines growing on Burgundy's Hill of Corton. We also, just for fun, analysed some supermarket Chardonnay wines from Argentina's Mendoza and Australia's Barossa. All the analysed values for iodine were absolutely minuscule. But what was striking – OK, surprising – was that the very lowest values came from the Chablis samples and the highest from the Barossa. Go figure!

Minerality

And then there's the big one. Apparently the most commonly used wine descriptor these days: the conundrum that is minerality. There have been conferences, masterclasses and workshops devoted to minerality in wine, countless scientific papers, magazine articles and blogs, and still they keep on coming. What is it about this topic that continues to fascinate? Who knows, but maybe it's to do with a melding of the pragmatic usefulness of the word, evocative of the much-loved link between wine and the land, with the continuing lack of consensus on what the term actually means. Minerality can mean what you want it to mean!

For starters, if minerality was simply due to the mineral nutrients in the wine, then it would be straightforward for a winemaker to manipulate things, say by enhancing minerality by simply adding more minerals. It doesn't work. For some, minerality may be unknowingly triggered by something in the wine, sulphur compounds or acidity perhaps, and hence an example of my second type of metaphor. For others, it's in my third group, an allusion to actual geological minerals even though they have no taste or smell.

The word appeared just a few decades ago, suddenly, like a new meteor in the night sky, and the wine world noticed. Wines were soon no longer described as steely, lean, austere and the like but mineral. At first it was restricted to cool-climate white wines like Chablis and German Rieslings but soon – anything. In a current magazine, for instance, an article reviewing 2021 Napa Cabernets, mentions mineral/minerality 34 times, together with specific minerals: graphite-iron minerality, saline minerals, volcanic minerals, etc.

But in view of the blur surrounding the term it's hardly surprising that science has found it, well, challenging to identify what in the wine triggers the sensation.

I'm not aware of anyone today disputing that minerality has to be metaphorical, but even so it seems that because of the very word itself, making some connection with the ground is irresistible. Thus, mention of minerality is often accompanied by something about the soils – the presence of stones, for

example, or deep roots. (Curiously though, I see in a current listing of Chablis wines that Grand Cru Vaudésir has 'impressive minerality' because 'there is so little soil'). Stones exist because they have resisted weathering – they are inert. They affect the physical properties of the soil but they can't give anything to the vine roots. They can't give minerality. Deep roots are seeking water; there are few available nutrients down in the sub-soil (by definition, little weathered) and there's certainly nothing magical down there.

Comparisons are sometimes made between wine minerality and the taste of bottled water. Most such water, however, has been drawn directly from the ground, where normally it will have been held for long periods (the average residence time in the UK is over a century), and in immediate contact with the host aquifer. So anything that's remotely soluble is taken in by the water, unlike the selective, regulated uptake of nutrient elements by vine roots.

Consequently, the mineral concentrations in bottled water, which obviously lacks the powerful aromatic compounds that give wine its flavour, are typically greater than those found in wine. And crucially, unlike in wine they can include plentiful chloride, sulphate, carbonate and bicarbonate, all of which demonstrably

contribute to taste and mouthfeel. Indeed, this is why such things are so important in the brewing industry (explored in Chapter 17).

In the wine world, however, minerality and similar 'geological' words used in tasting remain nebulous. It's almost as though they have taken on a life of their own! None of this is to suggest it is mistaken to use geological words in flavour descriptions, or that it is unscientific. Indeed, science likes to employ metaphors: the 'tree of life', for example, 'soil health', the 'greenhouse effect', 'genetic fingerprinting', etc.

But we have to be mindful that with minerality and the like we are not just making comparisons, we are imagining things, mentally conjuring what rocks and minerals *ought* to taste like. It's not wrong. After all, we need all the words we can get in our tasting lexicon. Putting our taste perceptions into words: that was never simple.

ABOVE LEFT: Sauvignon Blanc vines with roots exploiting nearly vertical weaknesses in the slate bedrock, Michi Lorenz Winery, Kitzeck im Sausal, Austria. ABOVE: River-rolled cobbles or '*galettes*' at Château la Voulte Gasparets, Boutenac-Corbières, France.

4. Granite: Rock of Grandeur, Rock of Ages

Granite: Rock of Grandeur, Rock of Ages

There's a Granite State (New Hampshire, USA) and a Granite Kingdom (Cornwall, England). There are Granite Cities (such as Aberdeen in Scotland and Saint Cloud in Minnesota); Granite City in Illinois got its name from making special kitchenware, popular because it was made to look like granite. Wall paint comes in colors such as Granite Rose, Warm Granite and Granite Fog. The word is, of course, just one of the hundreds of names for different kinds of rock, so how is it that this particular one has become so esteemed and so evocative?

Strength with polish

Well, perhaps it's something to do with the many admired uses of granite, which arise from its exceptional physical properties. One example is the rock's renowned strength. Thus granite has been used for grand buildings, such as the vast Palacio de San Lorenzo, near Madrid, and Aberdeen's majestic Marischal College. It's used for bridges: in London there's Tower Bridge, Waterloo Bridge, the old London Bridge now beside Lake Havasu in Arizona and, for that matter, the strengthening embankments of London's river. The iconic, isolated lighthouses perched on rocky reefs out in the waters surrounding the British Isles – Skerryvore, Bell Rock, Eddystone, etc – are incessantly pounded by the waves yet they have prevailed since Victorian times: they are built of granite.

Then there's its handsome polishability. The rock can take a high degree of polish and retain it for many years. As many a kitchen shows: 'The sheen of our polished granite countertops adds elegance to any space'; 'The lustrous beauty will never wear'. And certain granites can take an exceptionally fine finish. Out in the Firth of Clyde in Scotland's southwest lies the isolated granite lump of Ailsa Craig, from where, curiously, nearly all of the world's curling stones come. Apparently the Ailsa rock has the perfect weight and polish for gliding smoothly over ice.

But what of granite in the world of wine? Enthusiasts may think of the great sites of France's Northern Rhône such as Cornas or the hill of Hermitage, or perhaps Australia's Granite Belt wine country, though, unsurprisingly seeing as granite is one of the most abundant rocks in the outer part of the Earth, vineyards sited on granite are widespread. But wait: vines cannot grow on actual rock but only in humus-bearing detritus derived from it. So if granite is so famously strong and indurate how come it breaks down to give soil?

All things must pass

It's all a matter of time. Most of the granite monuments we think of are only a few hundred years old or less – a mere tick in geological time – yet close-up they commonly show signs of decay. Even the outstandingly well-preserved 3,000-year-old 'Cleopatra's Needles' (there are actually three of them, in Paris, London and New York) are now worryingly clearly deteriorating. In other words, although granite is robust, given time it isn't as immutable as we like to think. So how does this decay come about?

Around 25 percent of the mineral make-up of granite is hard, glassy quartz, and less than 10 percent consists of constituents such as mica. The major component is the mineral feldspar, mainly potassium feldspar, and this is the Achilles heel, so to speak, of the rock's longevity. Water, and especially slightly acid water, slowly reacts with the mineral to form kaolin clay. (It's why using mild acids such as lemon juice and vinegar to clean a granite worktop can harm its surface.) So over geological time this can lead to accumulations of kaolin-dominated soil. For instance, at Larnage, in the Crôzes-Hermitage region, the kaolin soils derived from the local granite not only produce acclaimed Syrah wines but are sufficiently thick to be quarried, for the famous Le Panyol clay ovens. On South Africa's Cape Peninsula there are old quarries in thick kaolin, at Noordhoek and Fish Hoek, just south of the historic Constantia and Steenberg wine estates.

This chemical action of water is usually accompanied by physical breaking of the granite, for instance by repeated heat-driven expansion and contraction of the rock, and this acts to disintegrate the granite as a whole. So although granite is the very archetype of ruggedness and durability – as in 'Arnold Schwarzenegger's jaw is a block of granite'; the 'granite-hewn colossus in Finnish culture' that is Sibelius – in the fullness of time it breaks down just like all other rocks.

Colour

The resulting detritus, moistened and mixed with decayed organic matter, is what we refer to as a granite soil. It's usually dominated by grains of hard, inert quartz though some of the feldspar might survive temporarily as blocky fragments. And because quartz is glassy white and feldspar usually a pale pink or grey, most granite soils are pale-coloured.

Drive along the IP3 up the Dão Valley from Coimbra, Portugal, or take Route 79 into the hills east of Temecula, California; strike east on any of the roads leading from South Africa's R44 towards Helderberg, Stellenboschberg or Simonsberg: the sandy gravelly soils all around are an arresting salmon-pink, because of the rosy potassium feldspar in the underlying granite. On the other hand, the whitish feldspars of the Sierra Nevada granite in central California tend to make the soils of the Zinfandel country of the Sierra Foothills a pale ashy grey.

Besides the quartz and feldspar, some granites have small amounts of silvery mica, making the soils sparkle in the sunlight. (And moonlight, for rumour has it that some growers in Portugal's Douro area used to surreptitiously gauge the granite content of their soils at night.) Or there may be various dark-coloured minerals. The soils of the Jeanraude vineyard of Condrieu tend to be pale in colour, but those patches that overlie a granite bedrock with black mica are noticeably darker. The Ajaccio wine region of western Corsica is dominated by granites with slightly differing mineral constituents, with resulting differences in colour and resistance to erosion. The escarpments around Petreto-Bicchisano, for example, on the road from Ajaccio to Bonifacio, are due to such differing granites.

Incidentally, away from geology, many hard, polishable rocks are called 'granite' irrespective of their colour or constituent minerals. For example, to a geologist the 'Ingleton Granite' being quarried in England's Yorkshire Dales is an impure sandstone and is not a granite, and neither is the celebrated 'black granite' of the poignant Vietnam Veterans' Memorial in Washington DC. (It's a rock properly called gabbro.)

OPPOSITE: The pale grey, massive granite at Mount Rushmore, South Dakota, USA, into which the faces of four presidents have been carved.

The water flows

In general, two features of granite soils are of most importance to grapevines. First is their free drainage and here's where the quartz constituent becomes particularly significant. It tends to persist as loosely packed rigid particles, and together with any surviving pieces of feldspar this leads to the open sandy soils typical of granite sites.

Growers prize the resulting free drainage. Indeed, because of it one grower in Virginia, USA, regards his granite soils as 'a buried *treasure* chest for making great wine' and another in the Maldonado region of Uruguay calls it his 'secret weapon'. According to a grower in Meerlust, South Africa, 'among the clay soils for our Cabernet the well-drained granite soils are something of an El Dorado'. (In that same part of the world, this free-draining property seems to be why elephants move to areas with granite soils in the wet season.)

Through being drier, these soils have the advantage of being relatively warm, which decreases the risk of root rot, and of encouraging beneficial, water-seeking deep roots. On the other hand though, lack of water retention can be a problem. In Alsace, warmer granite-based sites such as Brand (Turckheim) and Schlossberg (Kientzheim/Kaysersberg) can suffer in drier years relative to cooler, moister ones such as Sommerberg (Niedermorschwihr) and Wineck-Schlossberg (Katzenthal). The size of the effect depends on exactly how much clay has developed from the granite. Some growers on the southern Cape Peninsula of South Africa believe that were it not for all the kaolin, once mined nearby, their granite soils would be excessively drained and just too dry.

The nutrition game

The other key characteristic of granite soils is their low fertility. While this would dismay a farmer seeking maximum crop yields it is practically undeniable in viticulture that vines with low nutrition will put their energies not into excessive leaf growth but into producing smaller grapes with concentrated flavours. The micas and other minority constituents of granite can usually just about provide for the modest nutrition needed to do this, but not much more. But the absence of calcium carbonate in granite results in acid soils that can over-restrict the availability of certain nutrients, such as magnesium and phosphorus. That's why some growers, such as around Darling, Paarl and Stellenbosch in South Africa's Western Cape, have to routinely incorporate lime in their granite soils.

No nutrition comes from the quartz in the soils, it being just silicon and oxygen (silica), or from kaolin, being just silica plus aluminium, none of which are needed by the vines. However, as the feldspar alters to kaolin, potassium is released and this can be a big nutritional player. Granite soils typically are rich in potassium.

This potentially could lead to nutrient imbalance, but it isn't the total potassium that matters but the proportion accessible to the vine roots. And this could be surprisingly small. A study of the granite-derived, potassium-rich soils of Lodi, California, found that over 90 percent of the potassium is locked up in the unweathered crystals of feldspar and wholly unavailable for nutrition. And of the remaining 10 percent or so, most is caught *inside* the kaolin that coats the decaying feldspar grains. Only potassium on the surface of the clay, which is less than two percent of the total potassium in the Lodi soils, is accessible to the vines. Even smaller values have been measured in places elsewhere.

Despite this, the vines' absorption-selectivity mechanisms may fail and more potassium is taken in than needed, which can be problematic. Any excess amounts may eventually reach the grape juice and there it will react preferentially with tartaric acid, to produce insoluble tartrate crystals. I've seen piles of purple crystals outside the South African winery of Boschendal (Stellenbosch), nestling on the granite soils of the mighty Simonsberg. They're accumulations of 'wine diamonds', those little crystals we sometimes see at the bottom of bottles and corks, stained purple in the case of red wines. They're tasteless but the wine may have become flabby and unstable as a result of the tartaric acid content having being lowered. In short, granite soils are generally regarded as desirable for producing quality wine but the grower has to keep his eye on them.

A massive benefit

The striking feature of a natural granite mass is its, well, lack of features. Think of huge rock faces like Mount Rushmore (South Dakota) or Yosemite's El Capitan (California): there are none of the weakness planes of rocks such as slate or schist and none of the stratification of sandstone or shale. This is because the granite formed from a large volume of molten rock that slowly cooled and allowed crystals to solidify into a homogeneous, uniform mass. It all happened down below the surface of the Earth, typically at depths of tens of kilometres or so. Then as through time Earth's internal processes slowly forced the cooled rock upwards and erosion gradually removed the overlying material, the granite became visible.

Geologists refer to this featureless appearance as 'massive', and it gives the rock a certain majesty. Thus 'Beethoven's Fifth symphony has a granite-like, statuesque grandeur'. And it allows remarkably large blocks to be quarried: the ancient Egyptians utilized it for their great temples and obelisks, still some of the largest stone monuments ever fashioned. All this, together with its strength, has given granite a reputation of unshakeability. (Chairman Mao's tomb is of granite 'as unshakeable as Mao's self-belief in his brutish doctrines'; Donald Trump is said to have 'a granite confidence in human corruptibility'). This perception of solidity has long made the rock a favourite material for prestigious major buildings – and to give banks an indestructible look.

An important reason why port producers along the Douro in Portugal have traditionally avoided the granite there is its massive nature. Water is crucial in this arid area (there are cacti in some vineyards) but unlike the nearby schists that allow the winter rainfall to percolate down their vertical planes of weakness and be stored for the torrid summer, the granite gives no such opportunity. In fact, the vine roots can barely penetrate the featureless granite.

There's a great deal of granite in Portugal. It's seen in winery buildings in Estremadura (Lisboa) in the south all the way up to the Vinho Verde region in the north, for example in tanks for fermenting and maturing the wines. Granite tanks are right now newly fashionable – away from Portugal they've recently been installed in, for example, the Torres winery in Spain's Pazo Torre Penelas (Rías Baixas), the Schmitges winery in Erden, Mosel, and the Giesen winery in New Zealand's Marlborough – but nowhere is there a more famous example of granite being used inside wineries than the *lagares* of the Douro region. Surely a classic image of the traditional wine world is a throng of exuberant souls dancing on grapes, with purple juice dribbling down their bare legs and feet. The ritual still continues in some places, almost always carried out in stone troughs, usually of granite.

Blocks of rock were quarried for this purpose in places like Vila Pouca de Aguiar, Portugal's self-styled 'granite capital', where the granite is sufficiently massive for slabs to be sawn from it and formed into rectangular tubs. (In Vale de Mendiz is a circular example, its builder apparently trying to avoid the stompers hurting their toes in square corners.) Granite is now rarely produced for new *lagares*; those being used today are centuries old, once hauled to the Douro area on ox carts.

Worlds of granite

Huge volumes of granite, having been formed at depth, are now at the Earth's surface. They dominate the mountains that parallel the Pacific coast of the Americas all the way from the Rockies of British Columbia down to Patagonia, yielding important vineyard soils. When you stand among the vines on the flat floor of the Elqui Valley in Chile, those soaring mountains all around you are carved in granite. At the experimental vineyard way up in the Peruvian Andes, near Machu Picchu, is the ancient Inca citadel with its stone buildings clustered beneath the towering peaks, and all the rock you see – including the buildings – is granite. The rock accounts for the mountains around Stellenbosch and Franschoek in the Western Cape of South Africa, and from there stretches northwest for 200 kilometres (125 miles) through Darling and Swartland. In Australia, granite around Armidale in New South Wales continues all the way to the Granite Belt wineries of Queensland, a distance of more than 250 kilometres (155 miles).

For some wine enthusiasts, archetypal granite country may be Beaujolais, although actually rather less than a quarter of the region as a whole has granite bedrock. It's in the 10 Beaujolais Cru areas where the rock does come to the fore. Chiroubles, for example, is entirely granite. And it's tough stuff. In places where the overlying crumbly soil is particularly thin growers have had to use drills and crowbars to loosen the substrate for the roots. Pink feldspar is easily visible in the rubbly vineyard stones. The upper reaches of the twisting D86 and D119 roads that climb above Chiroubles village give glorious views across the terraced vineyards, strikingly buttressed in places with piles of the local pink rock.

Moulin-à-Vent is largely underlain by granite, though by no means wholly. Some of the most highly regarded *lieux-dits* (eg Le Clos, Le Carquelin, Champ de Cour, Les Thorins) are granitic, and flank the old windmill that gives the commune its name. The mill, used centuries ago for producing flour, is now much restored and its walls stuccoed but, unsurprisingly, its walls are made of granite. Down in the Côte de Brouilly Cru the celebrated rock so conspicuous all around is often called granite, but wrongly. It's a dark blue-grey colour, unlike granite, and much used in the local buildings, where a close look shows little in the way of quartz and potassium feldspar: technically it's a rock called diorite.

Around Clisson in France's Muscadet country the bedrock is granite and its extent defines the Clisson Cru Communal. Granite rules here, not only for the 20 or so wine producers but for the town itself. Its bridges, churches and houses are all built of the local rock, as are the walls of the wonderful 15th century

timber-framed market hall. And granite is still relevant today, with disused quarries in the area providing top-notch technical climbing walls. Can it be coincidence that a three-day music festival takes place here each June: the heavy metal 'Hellfest' featuring, you guessed it, top rock bands?

The immovable rock moves

In some vineyard areas with granite bedrock the overlying soil is derived directly from underneath, apart perhaps from some slipping down hillslopes.

This is the case, for instance, in parts of Alsace. The vineyards there stretch along the steep hillsides caught in the intricate fault zone that separates the easterly Rhine Valley from the Vosges to the west, upstanding partly because of their tough granitic foundation. A few vineyard sites have soils largely of granite scree tumbled down from the adjacent Vosges.

Examples are at Brandt and Wihr-au-Val in the Vallée de Münster; a Grand Cru wine from Sommerberg in Niedermorschwihr is simply called 'Granite'. There are several different granite bodies involved and so the soil colours vary with the mineral content of the parent rock: at Dambach the material is a buff-pink, as in the town walls, whereas the ruddy feldspars at Châtenois give a deeper brown-red. Many of these hillslope sites are crowned by old castle ruins, perhaps most famously at Schlossberg, Alsace's largest Grand Cru, overlooked by the brooding granite walls of Kaysersberg Castle.

LEFT: Granite soils at Franschoek, Western Cape, South Africa. Their pale colour is due mainly to the quantities of feldspar and quartz in the soil.

On the other side of the world, the vineyard slopes of the granitic Strathbogie Ranges, northeast of Melbourne, are less steep than Alsace but nevertheless much of the granite soil has slipped downhill. The vineyards have soil peppered with large granite boulders that have been transported downslope to their present position, but that the bedrock is granite is shown by the imposing crags all around. This was Ned Kelly country. The outlaw spent 16 months on the run here, presumably utilizing the cover provided by the bluffs and boulders of these rugged granite hills.

There are vineyard areas with soils in which granite is important even though the bedrock may consist of something else. In most cases the granite has been brought to the site as pebbles and boulders by rivers and nowhere is this more apparent than in the great valleys of central Chile. Charles Darwin realized this, as back in 1834 he recorded that he'd found 'numerous large blocks of granite' along the Maipo and Cachapoal rivers even though 'granite does not occur within a distance of 20 leagues'.

The granite of these alluvial pebbles continues to weather in place: they can be sufficiently crumbly for vine roots to be able to grow through them, to reach the more clayey material below. Some pebbles even fall apart when picked up. One producer believes that the degree of the weathering affects the wine: more intact granite pebbles give 'medium volume wines, with good weight on the palate, and well balanced' but increasing degradation leads to wines with 'greater structure, a broader and heavier feel in the mouth, and smooth tannins'.

Over on the Argentinian side of the Andes there is much more input into the alluvial gravels from nearby volcanoes such as Tupungatito; consequently many of the pebbles are made of rocks like the eponymous andesite and basalt.[1] In places, however, granite is significant. In Gualtallary, for example, in the north of the Uco Valley, the alluvial soils of the Las Tunas River contain granite boulders reaching the size of cannonballs.

Of granite and grapes

I often read that particular kinds of grapevine 'prefer' specific geological soils and, conversely, that certain kinds of geology are best for particular cultivars. I've seen the notion presented in tables, with stars for the favoured combinations, like

1 Andesite – first described in and named after the Andes Mountains – and basalt are both volcanic rocks, once molten like granite but with much finer constituent minerals and lacking quartz.

Riesling and slate, Cabernet and gravel, Chardonnay and limestone and so on. And granite? Well, Gamay and Syrah. Thus – so the idea goes – for quality wines from these two cultivars you need granite soils.

But surely these associations are simply reflections of certain classic European sites? It seems to me that which particular cultivar grows well at a site depends on the mix of all the relevant factors and not just the geology (not to mention the kind of rootstock onto which the cultivar has been grafted). Clearly Gamay thrives in the granite soils of Beaujolais and similarly Syrah in the Northern Rhône but both cultivars also flourish and yield superlative wines in the other soils of those two areas. In the Rhône, for example, the Côte Rôtie is not granite and the hill of Hermitage has soils other than granite. And conversely, Condrieu is composed of granite but isn't planted with Syrah.

It's similarly diverse elsewhere in the world. Syrah vines obviously thrive in the granite soils of, say, South Africa's Swartland and Paarl, but they also flourish in Australia's Barossa, New Zealand's Hawke's Bay and Washington State's Yakima Valley, none of which involve granite. The Paarl vineyards also grow successful Pinotage, Petite Sirah, Viognier and Grenache. Wineries in the Sierra Foothills of California are proud of their granitic soils but here it's Zinfandel that rules, together with other cultivars that yield outstanding wines of their type: Barbera, Petite Sirah, Mourvèdre and many more.

Granite and wine

In a given area with granite soils the interaction of terroir factors may bring some character to the finished wine. As just one of numerous instances, a grower in Alsace produces Riesling wines from granite soils and from limestone. For him the granite wines are 'more aromatic, lemony, limey, friendlier', they don't 'have the tannic structure of wine from limestone'. Consequently, and somewhat ironically in view of the strength of the rock, he says 'granite gives a more fragile wine'.

Presumably such differences come about through the interplay of the water and nutrient aspects discussed earlier, but probably together with many other factors. Localized wild yeasts for instance, both in the air and in the soil, may be

OVERLEAF: Granite soils at Cornas, Northern Rhône, France, with crags of granite bedrock in the middle ground. The southerly aspect of the slopes gives good sun exposure while their steepness together with the sandy nature of the soil give excellent drainage.

playing a role and there is research suggesting that the mere presence of pebbles in a soil can affect them. The stones exist because they have so far resisted erosion and so obdurate granite is a common constituent; the effect, though, arises from conserving tiny amounts of moisture below the pebbles and isn't anything to do with their being made of granite.

Extrapolating these effects from a particular area to generalizations is hard to justify scientifically, although there are those who believe a granite origin can be detected in wines from anywhere. Granite is made of just the same kinds of minerals that make other rocks, albeit in particular proportions and fitted together in a way that reflects their molten origin. Chemical analyses of a typical shale, schist and granite, for example, are remarkably similar. In other words, although there are claims like 'granite soils give a special energy and vibrancy to the wines', granite has no special ingredient to bring to wine. Also, it is noticeable how inconsistent wider claims about the effect of granite on wine are.

For example, white wines from granite soils have been said to be 'tight, edgy wines which are accessible young' but also 'all about substance, breadth and longevity'. For some the wines show 'opulent aromatics, with tropical pineapple' yet for others they're 'exceptionally mineral with an edgy sensation of salinity'. With reds, while for some 'granite gives more bass notes, power and viscosity', for others the wines are 'floral, fragrant with high tension'. Texturally, some say that 'granite soils give an airy texture'; others sense 'dense strongly textured wines'.

Even more problematic, scientifically, is the notion that granite can actually be tasted in the finished wine. Literally. I see in tasting notes things like: 'you can really taste the granite'; 'the granite definitely comes out in the wine'; 'it has warm, granite tones, with riveting granite aromas'. Such phrases may help convey a sensation but they have to be some sort of creative metaphor. We can only taste liquids. As mentioned, there are solids that are so very soluble that in our mouths they instantly start to dissolve and so we can taste them – salt and sugar being the obvious examples. But famously obdurate granite: it's just about as far as you can get from this on the solubility scale. It simply has no taste.

Rock of reverence

I see wine labels describing the contents of a bottle as being 'from granite soil' or something similar. For the reasons just discussed, for me this conveys little of what I might expect from the wine but I do like seeing such information. Wine is, of course, so much more than the liquid in the glass, so knowing about its background geology helps round out the picture and give context.

It might be enriching just to know that so noble a rock has been involved. Noble because, for instance in the US, according to the US Geological Survey, granite is America's most widely used stone for grand public buildings and monuments. The Washington Monument in Washington DC is largely composed of granite, as is Maine's State House and the capitol buildings of Texas, Arizona and Wisconsin; granite has been said to be integral to the founding of New York City. Or the phrase may conjure up images of granite landmarks associated with vineyards, perhaps South Africa's majestic Paarlberg, the Texas Hill Country's Enchanted Rock or the peaks flanking Arizona's Chino Valley.

Knowing granite was involved might add to a spiritual dimension of the wine as for some it's a blessed rock. The country of Zimbabwe takes its name from the sacred Dzimba Dzemabwe granite monuments that are central to its culture. The granite that today encircles the ski slopes of Utah's Alta and Snowbird was ordained by Brigham Young to be used for his definitive new temple in Salt Lake City. Aboriginal Australians still venerate granite as a 'stone of protection'; indeed, some western New Age folk believe granite to be a 'guardian stone'. Thus one website offers pieces for sale: 'an ideal gift for people who have high-risk jobs, like military personnel'.

Be all that as it may, granite is certainly a special rock. It's historic. It was a hallmark of Imperial Rome: thousands of granite columns spiked the skyline and adorned the city's great buildings, such as the Pantheon, the Forum of Trajan and the Baths of Caracalla. In a way, the USA was founded on granite in that the rock constitutes the nation's most historic monument, Plymouth Rock in Massachusetts, supposedly the landfall of the founding Pilgrim Fathers. Surely our enjoyment of a fine wine is enhanced if we know that back in the vineyard the ground involved this noble rock, this rock of pedigree and grandeur, a rock of ages?

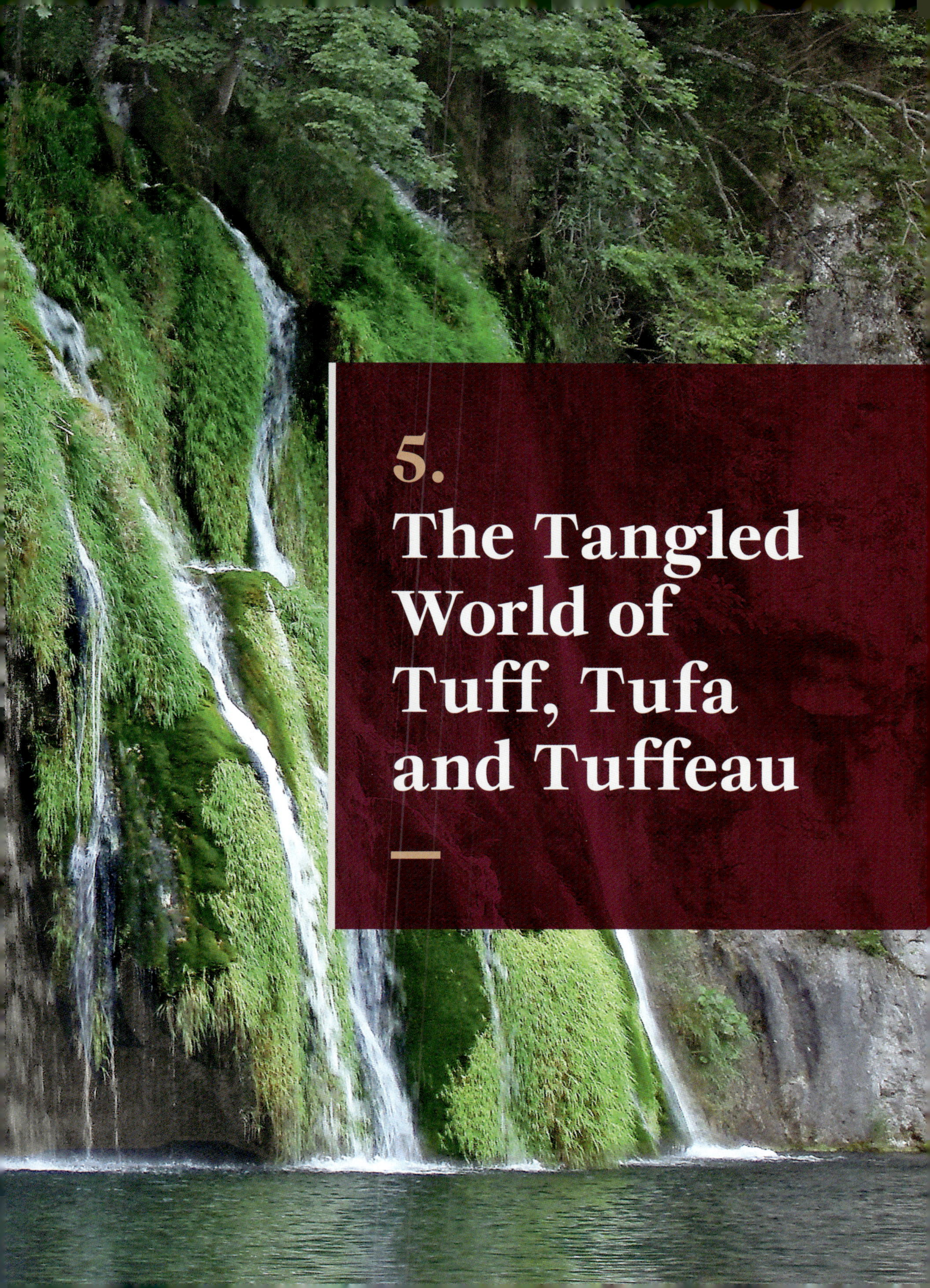

5. The Tangled World of Tuff, Tufa and Tuffeau

The Tangled World of Tuff, Tufa and Tuffeau

Tangled indeed. And to these tricky 't' words I could add tufo, tuf, tufi, tofa, tuft – and others. We are talking, of course, about names for geological materials, names that, because most of them occur in vineyards around the world, appear in the literature on wine. You may have seen the kind of thing: 'the cellars, as is typical of the Eger region, are cut into the soft tufa'; 'Saumur – the capital of tuff'; 'the soils are primarily clay above volcanic tuft'; 'the ground consists of calcareous tuff'; 'the Classico zone of Orvieto consists of tufo, similar to the tuffeau of parts of the Loire'.

So what's the problem? Well, to a geologist all those quotations in one way and another are simply wrong. In geology, each term refers to a material of a specific nature, with a meaning that doesn't fit with those quotations. Outside geology, however, these 't' words tend to be used loosely, even synonymously and interchangeably. Thus, there's a difficulty: their meaning in wine writings is often unclear. But neither are things tangle-free within geology; for instance, there's another rock in the frame, the celebrated travertine, which geologically is closely related to tufa but awkward to distinguish. More of that later.

The root of the confusion lies back in classical Roman times. When the architects and builders of the infant empire began to construct grand buildings worthy of its new status, they found to hand, right there in the capital, material that was easy to extract in blocks well suited to the monumental style they had their sights on. They called that material *tophus*, sometime *tofus*. We now know that in Rome some of it consisted of rocks derived from the fine ejectamenta of the region's volcanoes, what geologists would now call tuff, and some came from

PREVIOUS PAGES: Tufa, in the UNESCO-listed waterfalls at Plitvice, Croatia. The moving water constantly erodes and continually deposits, so the shape of these cliffs is always changing.

matter precipitated from flowing springs, which includes the tufa known by modern-day geologists. It was all blurred together.

Then, as blocks of building stone became employed across the expanding empire, they too were referred to as *tophus*, irrespective of what the rock was. And to complicate things further for us today, the word became variously transmuted into the local vernaculars. Thus, for example, *le tuffe* evolved in French, and tuff in English. In Britain the word 'tuff' long held sway for any porous, softish building stone that wasn't obviously, say, limestone, sandstone or granite. By the 18th century, however, when the British stone dealers of the day learned that their Italian counterparts were using the word *tufa*, they took the opportunity to try to convince architects that this exotic-sounding word was preferable to old-fashioned, Anglo-Saxon sounding tuff. So now we had two words in English, tuff and tufa, being used for convenient building blocks irrespective of the kind of rock.

However, geologists saw during the 1800s, as their science was growing, that this was unhelpful to understanding and a consensus evolved that the name 'tuff' should be restricted to volcanic material. Consequently, for well over a century now, geologists have used tuff to refer only to volcanic rocks, and tufa for a calcareous rock formed by localized precipitation from cold groundwater. However, outside geology the lack of distinction still hangs on, and hence the confusion. Surely, if we are to understand vineyards, how rocks and soils might affect the growing vine and perhaps the wine that results, we have first to be reasonably systematic with the terms we are using. So here's an attempt to untangle the situation.

The Eternal City and its tuffs

The Romans were certainly not the first to use these kinds of stone for building, though the evidence from earlier times is slender. Rome is founded on volcanic tuff. The rock makes the fabled seven hills and by the sixth century BCE it was being quarried for the solid stone blocks that befitted buildings of the empire's growing status. Several writers mentioned it, including Strabo, Ovid and Virgil. Pliny remarked that this local *tophus* weathered badly and needed protection, for instance by applying a limewash.

Apparently it was the same in Carthage, North Africa, where the local *tophus* was so vulnerable to the salt air that it had to be coated with pitch. Thus, quipped Pliny, while in Rome pitch was used to seal the insides of wine amphorae, the

Carthaginians used 'pitch for their houses and lime for their wines'. But it was the great architect Vitruvius who documented in some detail the *tophus* found in and around Rome and its suitability for construction.

First though, let's be clear what modern science means by tuff. The molten rock sitting below an active volcano may be relatively thin and runny, such that if internal pressures force it to erupt it will quietly ooze out as the lava we often see in video clips of, say, Hawaii. But if its chemical composition makes it thick, viscous and gas-rich, it may break out with explosive force, resulting in a violent mixture of gases, lava and fragments of already solidified rock. Much of the coarser material will accumulate around the eruptive vent, within time building up a cone shape as in the popular image of a volcano.

Finer material will travel greater distances, either by racing across the land surface as incandescent clouds, or dispersing high in the air as fine ash. Eventually even the finest materials will come to a rest, typically settling in layers and, with time, hardening into rock. This is tuff. Incidentally, geologists have a word that covers the full range of debris ejected from volcanoes, and curiously enough it's another 't' word. Tephra, however, is related to the Greek word for ashes, *téphra*, rather than the *tophus* line, and it doesn't seem to have entered the populist confusion.

To be called tuff, the component particles have to be no larger than a couple of millimetres or so, and the whole mass must have been discharged from a volcano. The degree of layering varies, and its colours and textures depend on the mix of geological minerals involved. The minerals themselves are composed of silicon and oxygen held together by elements such as potassium, sodium, iron and magnesium: importantly, they are *silicate* minerals, the kind that make most rocks. In other words, *tuffs are volcanic and siliceous in composition*. And this, as we will see shortly, contrasts with the calcareous precipitates that make tufa. No volcano in the world erupts calcareous rocks.

Back to Vitruvius. He noted that stone used for the early buildings of Republican Rome was quarried from places more or less in the city, such as the red tuffs found just to the north at Fidenae (where, according to one account, 'tolerable wine' was made) and the moderately good (*temperatae*) stones of nearby Tivoli, some of which we would now call tufa. Then, as the Romans extended their influence over nearby lands, they were able to access more durable varieties of *tophus*, first from the Monti Sabatini volcanic deposits just to the north, and later the harder tuffs to the southeast that came from the Albani volcano.

Quarries remain at Albani today, for example at Tuscolo in the present-day Frascati DOC and Marino in the Colli Albani DOC. Incidentally, the most recent eruption of the Albani volcano was 36,000 years ago. That may seem an immense time to us, but we are reminded that it's geologically trivial by the recent spate of minor earthquakes in the region and the appearance in 2013 of a steam vent near Rome's Fiumicino airport.

Tuffs from further afield

The expanding Roman Empire soon took in Naples and its environs, so with all the products of Vesuvius and the other volcanic centres of the Campi Flegrei it's not surprising that tuffs were much utilized in this region. They are seen today in many of the city's old buildings, and they dominate the natural skyline. The conspicuous hill that is capped by the San Martino monastery, for instance, with a centuries-old vineyard clinging to its vertiginous slopes, consists largely of different kinds of tuff. Just to the west of Naples are numerous other volcanic centres, almost all with associated vineyards sited on tuffs, such as Averno, Astroni and Pozzuoli, and the islands of Ischia and Procida.

But just as in Rome, another kind of *tophus* – tufa – was also used for building. The ruins of Pompeii, for example, are chiefly made of tuff (albeit called in many accounts tufa) but the foundation blocks of some of the houses, together with the decorative stonework in their alcoves, are of true, geological tufa. For even in this highly volcanic area there are calcareous rocks around, for instance underground just to the east of Pompeii at Scafati and Mariconda, in the cliffs of Sorrento, and out beneath the sea to Capri. We shall see more of these in a moment.

Tuff is found across Campania but a noteworthy example, known now in its Italian form as *tufo*, occurs by the river Sabato about 10 kilometres (six miles) north of Avellino. Here, in about the 10th century, the underground workings of this quality building stone and associated sulphur deposits were greatly enlarged; consequently the nearby town grew, and eventually took its name from this desirable rock: Tufo. In turn the vines that had grown there for centuries absorbed the name, and so we now have the grape Greco di Tufo. It's an unusual instance of the contribution of geology to wine!

There's tuff in all the other Italian volcanic areas: Vulture, Soave and Etna, for example. It's important on volcanic islands such as Santorini, Madeira and the Canaries. France's Côtes du Forez in the volcanic Massif Central, Germany's

Kaiserstuhl in Baden-Württemberg, and Hungary's Eger and Tokaj regions all contain tuff, albeit of very varied kinds. The highlands bordering the Napa and Sonoma valleys in California contain a great deal of tuff, as does the northeast part of Australia's Barossa Valley and the area around Auckland, New Zealand. Some of the oldest vineyards we have, in Georgia, Armenia and Turkey, are sited on tuff.

But tufa is quite different

Having looked at the nature of tuff and examples of its occurrence, I turn now to another part of the original *tophus* in Rome – tufa. Despite having the same etymological origin, tufa is geologically quite different to tuff, being made of calcium carbonate, the mineral calcite. Hence, it's calcareous, it's a particular kind of limestone, and technically it's a precipitate: here's how that works.

Natural waters may have a substance happily dissolved in them but if they encounter some new circumstance, perhaps a drop in temperature or pressure, or the arrival of certain organisms, they may no longer be able to hold that substance

ABOVE: Soil at Tokaj, Hungary, composed of fragments of pale-coloured andesite tuff – originally the ash ejectamenta of volcanic explosions then lithified into solid rock.

in dissolved form. So it precipitates out, and we call the resulting solid a precipitate. This happens particularly with calcium carbonate because calcite is more soluble than the silicate minerals that make most rocks. But just how well it dissolves depends on things like the temperature and especially on the amount of carbon dioxide that may be in the water. Most natural waters contain some of this gas because it's all around us in the air, and microbes and other organisms living in and around the water have a lot to do with influencing its carbon dioxide content.

All this is why most caves are in limestone, the result of the constituent calcite having been dissolved away, but it's also why stalagmites and stalactites, made of calcium carbonate, may now be precipitating out. It's why objects hung up in so-called petrifying wells become 'turned to stone' or, more accurately, coated by calcite that's being precipitated. And it can happen around springs, as the underground water emerges and for one reason or another finds it can no longer sustain dissolved calcium carbonate. The precipitated calcite deposit is called tufa.

It's typically a softish, porous, pale-coloured rock, and often contains fossilized remnants of organisms, the very ones that may have had a role in the precipitation. More relevant for vineyards than localized mounds of tufa around a single spring are the more sheet-like occurrences that can cover substantial areas. For example, tufas can be precipitated from rivers, especially where there is limestone or other calcareous material in the vicinity, as happens in the Ebro and Duero valleys in Spain. Perhaps the world's most spectacular tufa deposits, albeit some way from important vineyard areas, are the river terraces and associated cascades at the UNESCO-listed Plitvice National Park in Croatia.

Sheets of tufa can also arise from lines of springs spread along hillsides. This happens in the Val di Noto, Sicily, where caves have been dug into the soft calcareous rock (but called by some guidebooks 'volcanic tufa'). The Aglianico grape is said to thrive in tufa, which is fairly widespread in Puglia, such as at Laterza and Castel del Monte. There are small plateaus made of tufa sheets in Tuscany's Val d'Elsa, southeast of Montepulciano, such as at Sarteano.

Tufa deposits are known around the world; some 500 sites have been noted in Hungary alone. Other European examples include: Touttour in Provence and Tournus in Burgundy; in Spain, at Priego in the Montilla-Moriles DO. A number of US states have tufa deposits: it forms in some upstate New York lakes, for example, and in Virginian rivers along the Eastern Appalachians, including the Shenandoah Valley AVA. In Australia, tufa deposits are scattered from Queensland right down to Victoria.

Does the tufa–tuff confusion matter?

It's unfortunate that tuff and tufa, both widespread in the wine world but geologically different materials, continue to be confused. But does it actually matter in practice? For many rock types, their differences have only subtle effects on grapevines, as most weather to provide well-drained soils yet offer some means of retaining sufficient water, and most can satisfy, together with the soil's humus, the relatively modest nutritional needs of vines. But the differences between tuff and tufa are more marked, and may at least affect a grower's methods.

Both kinds of materials are likely to give well-drained soils, stony in the case of tufa; tuffs weather to give water-holding clays, and a tufa bedrock may well be full of tiny fissures and holes that allow some water retention. But tufa, like other calcareous materials, yields soils that are distinctly alkaline. So a grower will have to think carefully about the most appropriate rootstocks. The availability of soil nutrients varies with pH (*see* Chapter 11): it's telling that there are instances of growers grinding up tuff and spreading it on tufa soils to enrich their nutrient content!

It's all very different with tuffs. Along with lavas, these porous volcanic materials are famous for yielding fertile soils not too long after an eruption. The warm humid conditions of tropical lands can foster luxuriant growth on relatively fresh volcanic soils, which are ideal for crops such as tea, bananas and pineapples. It's no accident that the world's great coffees are produced from volcanic areas.

But such conditions, even in a more Mediterranean climate, may not be good for grapevines: it's pretty much a standard refrain that a meagre nutrient availability is best for quality grapes. Although in the Naples region the Romans talked of the *Campania Felix* – the Fortunate Country – those vines on volcanic soils scrambling high over trees and yielding three grape harvests a year probably didn't lead to fine wine.

The differing colours of the soils may to some extent affect their thermal behaviour – some producers in Sicily's Val di Noto report a difference in grape ripening between vines on the white calcareous deposits and those on the adjacent dark volcanic materials. And presumably such influences on vine performance may affect in some complex way the taste of the eventual wine, though there's still little scientific substantiation of this. However, there is a growing belief that tuff bestows certain flavours on wine, part of the fashionable concept of 'volcanic wine'. (It's hard to see, though, how such a commonality can arise, given the great variability in physical and chemical properties of tuffs, let alone all the other kinds of volcanic rocks and soils.)

Similarly, although there are plenty of conflicting opinions, there is a widespread conviction that limestone, and therefore presumably tufa, somehow imbues wine with certain qualities (*see* Chapter 11). One commentator contrasts Sicily's Vittoria DOC wines produced from calcareous tufa ('lighter, fresher, more structured,' he says) with the more 'tannic, unctuous and mineral' wines produced on the nearby volcanic tuff (though he calls it tufa). Confusingly though, while growers with tuff soils assert that they bring greater minerality to their wines, producers with tufa and other limestones extol the enhanced minerality that their soils give.

Travertine, tuffeau and other 't' words

Having emphasized the differences between tuff and tufa, what about those other 't' words I mentioned at the outset? First, it's worth mentioning that *tufa* did exist as a Latin word but referred not to building stones but to the little feathers – tufts – that some soldiers wore in their helmets. As for *tufo*, the word that derived from *tophus,* it's used these days by Italian geologists as the word for tuff. But tufo is seen in English too, in populist travel accounts of Italy, but often with unclear meaning.

Sir William Hamilton, erstwhile British Ambassador to Naples (1764–1800) and married to Emma, Horatio Nelson's storied mistress, gave several descriptions of the geological materials around Naples. But curiously, in exactly the reverse meaning of today he defined tufa as a mix of volcanic ashes and pumice, and the porous stone associated with encrustations from water as 'tuf' (*sic*). Several Victorian topographical accounts refer to a rock called *tufi*, but these almost certainly are misprints.

But there's a 't' word to consider that's wholly bona fide. Travertine. It doesn't *sound* similar but it's so intimately related to tufa that geologists have long debated how best to define the difference. And it's a familiar material, creamy white and polished, with irregular dark holes – think of the façades of a well-known restaurant chain with golden arches. Most tile suppliers stock travertine. Many a grand building in Europe (visualize the shining white of Sacré-Coeur in Paris) and North America (New York's Pennsylvania Station; Chicago's Union Station) are at least faced with travertine. Mediterranean towns were built with it, such as the Italian Marche's Ascoli Piceno or Turkey's Pamukkale. You may have slipped on the shiny, shoe-buffed streets of towns such as Rovinj or Dubrovnik, paved with gleaming travertine.

Travertine, just like tufa, forms from the precipitation of calcite, however, it's usually associated with hot water. Greater amounts of calcium carbonate are dissolved, leading eventually to deposits of considerable thickness. The rock is normally denser than tufa and hence it takes a better polish. The warm waters also allow a lot of carbon dioxide to be dissolved, which on cooling comes out as a gas, which prompts the precipitation of calcite and together with decayed organisms gives rise to the irregular cavities and holes that are typical of much travertine.

Once again, the heartland is the area around classical Rome. The material was listed by Vitruvius as *Lapis Tiburtinus*, after Tibur, then a town about 25 kilometres (16 miles) northwest of Rome. It grew to become the town of Tivoli, these days a popular day trip for tourists visiting the capital. The *Tiburnus* stone became *tivertino* in Italian, at first, and then *travertino*. And then travertine in English.

The travertine quarries were active at the time of Vitruvius and they're still busy today. One modern author calculated that 100,000 tonnes of travertine were extracted from the quarries at Tivoli just for Rome's Colosseum! Later, Michelangelo used it to face St Peter's Basilica and its dome; Bernini used travertine from the nearby Guidonia quarries for the splendid colonnades that arc around St Peter's Square. Today, destinations for the rock are less rarefied: along with the quarries at Terme di San Giovanni near Siena, modern quarrymen are toiling to produce facing stone for the McDonald's restaurants of the world.

The name 'travertine', or its equivalent in other languages, appears on a number of wine labels, such as in Abruzzo's Colline Pescaresi and Tuscany's Grance Senesi and Monteregio, in Pouilly-Fumé and in Saint Pourçain-sur-Sioule in the Auvergne; it's the name of a winery in Australia's Hunter Valley. As with tufa, being calcareous, travertine will tend to yield well drained but alkaline soils, with all that means for vine nutrition.

So what's the geological difficulty with distinguishing travertine? It's the question of on what basis and where to 'draw the line' with tufa. Overall, travertine is precipitated from hotter waters, but if the material is ancient, to learn the water heat we have to interpret tricky analyses of isotopes and even then agree at what arbitrary temperature the terms should change.

The problem is well illustrated at one of the world's best-known travertine deposits, Mammoth Hot Springs in Yellowstone National Park, USA. They're

OPPOSITE: The calcium carbonate terraces at Mammoth Hot Springs, Yellowstone, Wyoming, USA. Travertine forms around hot springs when the temperature of the water is at least 70°C.

constantly changing, as any repeat visitors will notice, and from time to time some of the waters are cold, producing what by most definitions would be called tufa. As well as deposits that are 'in-between'. So what to call them? But that's a debate for geologists and, because all these materials weather to give pretty similar, alkaline vineyard soils, this distinction wouldn't seem to be important for vines.

Finally, there's a plethora of variations on these 't' words in the other European languages that don't need to concern us here, except one. The French word *tuffeau* is now thoroughly absorbed into English. The term applies to the layers of the softish, calcareous but somewhat sandy rocks that are found in southern Belgium and northern France, including the eastern Anjou and Touraine AOC areas (*see* Chapter 8). It is, of course, mentioned often in accounts of the vineyards and wines of those regions.

The lowest tuffeau strata, deposited in the warm shallow seas of the time, are referred to in French as the *Tuffeau Blanc* or *Tuffeau de Bourré*. It's a famed building material, used for Loire châteaux such as Chambord, Blois and Amboise, as well as grand buildings further afield, in Nantes and Rennes, for example.

Bourré is a little village along the river Cher, a tributary south of the Loire, right by the illustrious Château de Chenonceau. These days the town is a magnet for tourists curious about troglodytes and mushroom growing in the caves that the quarrying made available. The layers of rock hereabouts are a metre or two thick, ideal for extracting large blocks for building stone, which is still extracted today. Interestingly, in view of the current vogue for praising the mineral taste of some of the wines produced here and elsewhere, I must note that promotional material for this stone in the building trade extols its soft, warm, velvety texture, and emphasizes that it's 'not at all cold and mineral'.

The strata that were deposited on top of the white tuffeau, now eroded away around Bourré but easily seen in the cliffs along the Loire itself, say between Tours and Saumur, are a much more yellowy colour. Although essentially limestone, a third or more of the content consists of sandy quartz grains together with mica and a potassium-iron rich mineral called glauconite that weathers to give the ochrous look. The French term for these rocks is *Tuffeau Jaune*.

The point of giving these descriptions is to emphasize that tuffeau is a sedimentary rock, really extensive and stratified in layers owing to accumulations through time of submarine debris. It is not a chemical precipitate. Thus, although the word tuffeau may sound like tufa, the nature of the two materials and their geological origins are different: *tuffeau is not a synonym for tufa*.

The 'tangled world' today

I've explained how the confusion that surrounds these terms is historical, but presumably because of the similar sound of the words it persists today. As further examples of the tangle, Italy's Piemonte wine region is made of calcareous and other sedimentary rocks, but numerous accounts have it that the UNESCO-designated cellars in such places as Canelli and Monferrato are 'dug into tuff'. Conversely, in the volcanic hills of northern Hungary extensive underground cellars have been carved into the soft tuff that abounds there, in Eger, for instance. Jagged volcanic fragments are easily visible in the cellar walls yet in the publicity material of a prominent winery, in a leading travel guide, and throughout at least three books in English on Hungarian wine, the material is referred to as tufa.

Italy's Vignanello wines are produced from vines growing on tuff thrown out of the Vico volcano, but according to one account they grow in a 'land made of tufa'. In the village of Vignanello itself, the wine cellars are 'dug out of tufa stone'. The site of the Vulsini volcano is now occupied by the waters of Lake Bolsena, and important for the vineyards of the surrounding Est! Est!! Est!!! wine region, some descriptions say, is their 'tufa soils'.

Even more jumbled, the southern half of the Maremma Toscana DOC includes 'hillsides with a calcareous volcanic tufa soil structure'. English tourist literature on the region trumpets the 'Land of Tufa' centred on Pitigliano – the 'Queen of the Tufa Towns'. Its Bianco wine is 'typical of tufa soils', which are 'particularly suitable for white grapes'. (But the official description of the soils at Pitigliano is 'acid or neutral, and free of carbonates'). Ash from the Vulsini volcano got as far east as Orvieto, and the photogenic hilltop town apparently has something like 1,200 caves excavated into its foundation of soft tuff. But several wine guides call it tuffeau, the sandy limestone of the Loire.

So that's the problem. Unless we have prior knowledge, reading that a vineyard's soils 'consist of tufa' or that a wine was 'grown on tuff' doesn't guarantee what the geological material actually is, let alone what the wine might be like. And it gets yet more tangled! In some vineyard commentaries, tufa and travertine are called 'sinter', a term that geologists restrict to precipitations of *silica*. These are nothing like as widespread as the more soluble calcareous equivalents, but they're important in certain places and guess what… In addition to the word sinter, the material is given names such as silice, silcrete, silicium, silex and sarsen. It's a glimpse of a parallel world: that of confusing 's' words. A whole other story!

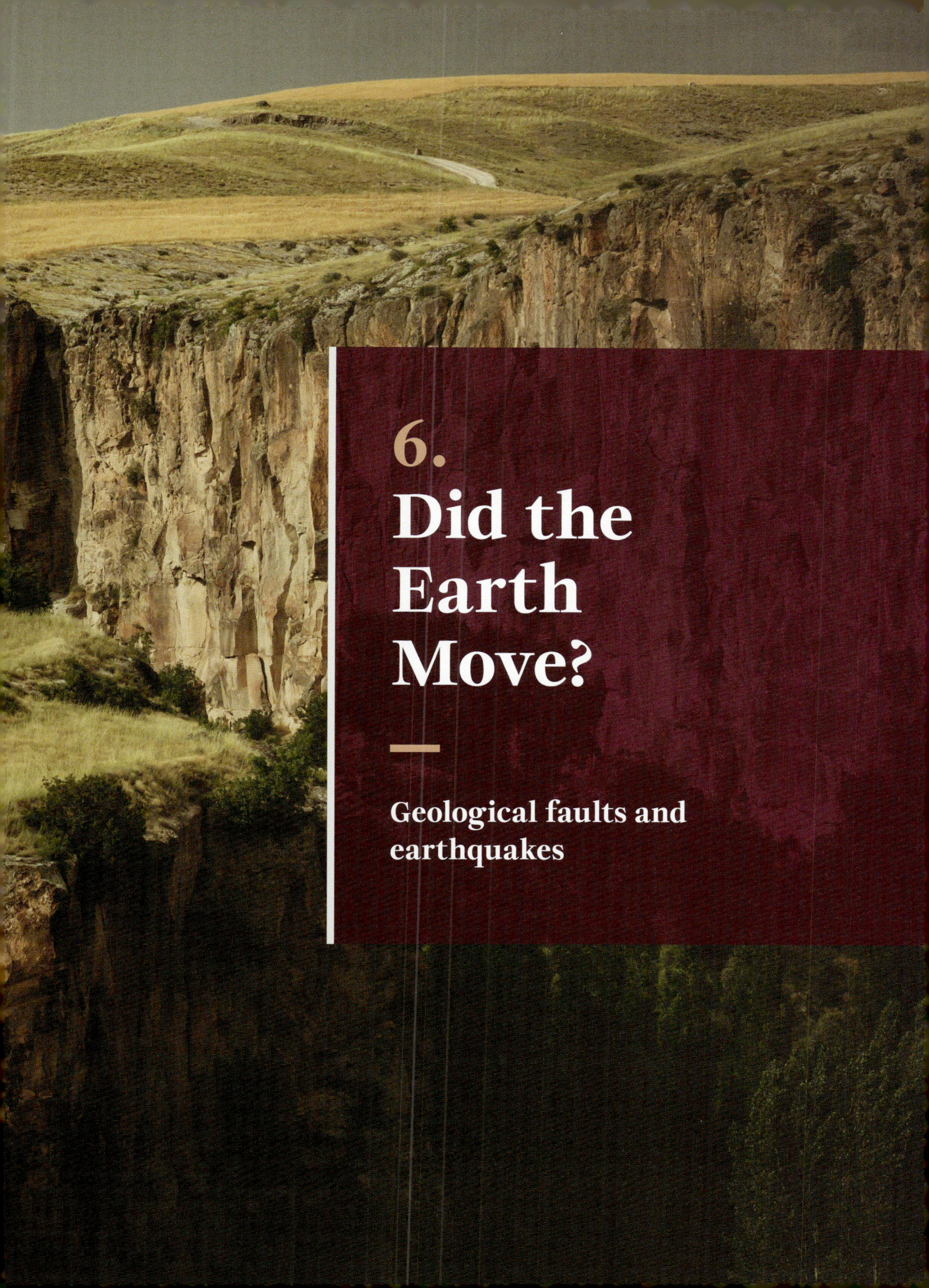

6. Did the Earth Move?

Geological faults and earthquakes

Did the Earth Move?

'But did thee feel the earth move?' Robert Jordan famously asks Maria in Hemingway's novel *For Whom the Bell Tolls*. Certainly, in many a vineyard the earth has moved – quite literally. For in the ground of a vineyard, particularly in the bedrock, ruptures are common, and along some of the breaks the materials have been shifted, physically moved. To geologists, fractures that show such visible displacement of the material either side are known as *faults*.

Well, that's the strict technical meaning, though the term has captured the public imagination as a metaphor. We talk of fault lines in commerce and politics, religion and culture: 'the survey indicates significant fault lines when it comes to marijuana policy'; 'women's rights and gender equality now constitute the world's fault lines'.

In this chapter, I will look at some aspects of geological faults relevant to viticulture, ending with some comments on perhaps the best-known geological fault of all, California's San Andreas Fault, which has inspired poetry, art, novels, films and music, as well as wine. But first, a few words on geological faults and what they are.

A quick look at geological faults

Just as remarking on faults in a wine can refer to some deficiency, in geology early usage of the word fault – from the Latin *fallita*, shortcoming – meant that something was lacking. When a miner was hacking along a coal seam or a vein of metal ore, he might find that it abruptly gave out, to be replaced by barren rock. So his target was suddenly lacking, and the miner himself was said to be 'at fault'.

But commonly the seam or vein was found again nearby, because it had been naturally shifted, and eventually the 'fault' became associated with displacement. Diagrams portrayed how a seams or veins – or geological strata – could variously be shifted, some showing the arrangements in exquisite detail. The emphasis, though, was on the *appearance* of the displaced strata; no one understood the cause.

As we shall see later with earthquakes, a certain mystery arose about geological faults and scientific explanation was a long time coming. Today, however, geologists know that stresses in the Earth – ultimately due to thermal variations within the planet – are continually waxing and waning, in different places at different times, and that sometimes they can become sufficient to fracture rocks and displace them.

Most ancient rocks have undergone a degree of faulting at some point in their history: it's actually a common phenomenon, although the displacement is typically small, just metres or so. The dislocation can, however, reach hundreds of kilometres in the world's big faults, such as in those special regions where major faults are active today and triggering headline-making earthquakes. Most of the these will have formed at the junctions where Earth's tectonic plates move past each other, and that's exactly what is happening today in places important in the wine world such as Italy, Greece, New Zealand, Chile and the USA.

The movement can be in any direction along a fault: sideways or 'up and down', and either or both sides of the fault may have moved. Only the smallest faults are likely to be single, clean fractures. So, although we usually picture faults as lines, in the ground they exist in three dimensions and usually have some

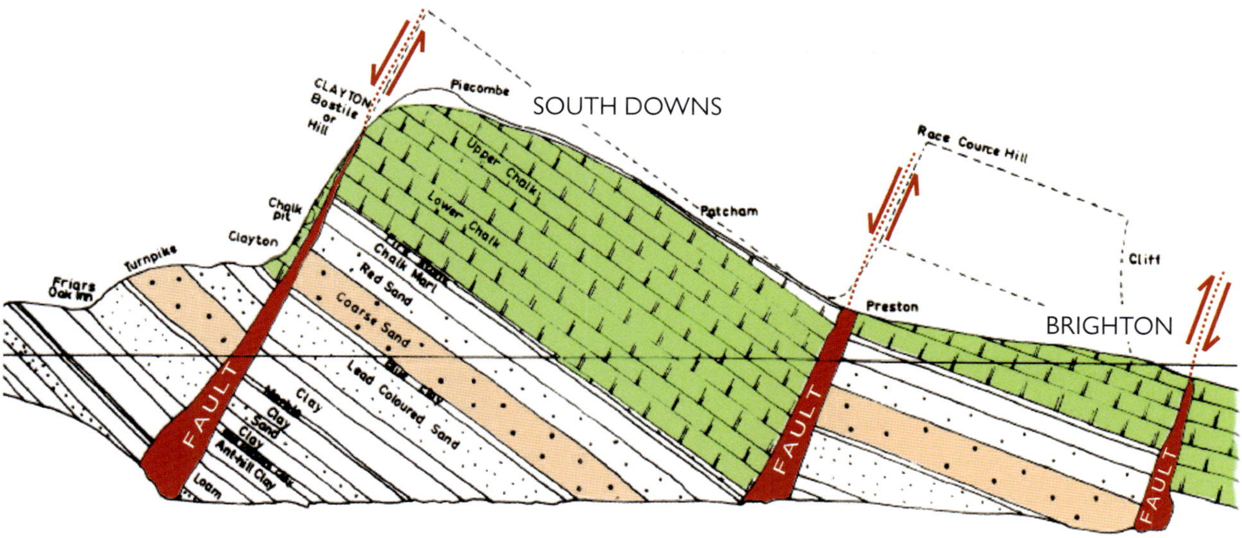

ABOVE: A cross-section across the South Downs of England, now a significant wine-producing area, drawn by John Farey in 1807 (colour added, for clarity). At that time, the idea of stratification in rocks, and its continuity being broken by faults, was in its infancy.

width of broken rock with further fractures splaying outwards into adjacent rocks. Hence geologists often talk of an overall 'fault system'.

If the Earth's stresses grow in a region that already has geological faults, then these weakened zones will be most vulnerable to further movement. The stresses may grow and eventually overcome the friction that holds the faulted rocks together, so that the rocks shift again and the accumulated energy is released – as heat, noise and the mechanical waves that we call earthquakes. Thus, fault zones can present ongoing earthquake hazards, potentially serious to wineries in such regions.

The water flows – or not

So now, following that glimpse of the nature of faults, I turn to what all this can mean for viticulture. First, the presence of geological faults in a vineyard can affect how water – so important for grape swelling – flows in the ground. For instance, in some fault zones the rocks break into jagged fragments with considerable space between them, which enhances water flow. In others, however, the rocks can be pulverized and finely ground such that the material becomes clay-like and acts to seal up the fault.

In Texas, USA, the Balcones Fault system curves from the Mexico border northeast to Dallas and has brought the tougher rocks that make the Texas Hill Country AVA next to the softer rocks that underlie the coastal plains. Water flow in the San Antonio-Austin-Waco line of wineries can depend on the nature of the local faults in the Balcones system: place names where flow has become localized here include San Pedro Springs, Comal Springs, San Marcos Springs, Barton Springs, Salado Springs and so on.

Jumping from soil to soil

Second, because at any place along a geological fault the rocks on opposing sides will have differentially moved, contrasting materials may have been brought next to each other. And if the overlying soil is derived from this bedrock, then it, too, will show sudden changes as a result of the movement. Although there are wineries that boast of the *consistency* of their vineyard ground – Cornas's Reynard vineyard, for instance is 'entirely pure granite, which gives the wine a purer taste with finesse and elegance'; the Fulldraw vineyard in Paso Robles has only 'pure white calcareous shale' such that 'the fruit you get is fresh and exciting' – it's more than likely that a winery with a vineyard straddling a fault or two will be trumpeting the *diversity* this brings.

Thus: in Sonoma's Rockpile AVA 'the Cemetery Vineyard straddles the Rodgers Creek Fault, giving it amazing diversity of soil composition'; Gigondas's Château de Saint Cosme is 'exceptionally located at the crossing of two geological faults, which is very rare. This gives us an extraordinary diversity of soils'; 'the Umpqua Valley AVA has an incredible 150 soil types because of its fault-lined teutonic (*sic*) geology'.

These enthusiastic claims of diversity can get quite competitive (and implausible): 'Burgundy is a land of faults, and the diversity of vineyard soils has no equivalent anywhere else in France'; 'The faults in Napa Valley give a spectacular diversity of soil...more than in the whole of France'; 'Because of the San Andreas Fault system, there is a greater variety of soil in Sonoma County that in all of Western Europe'.

But in any case, why might such soil variations matter? Well, for one thing they may influence a grower's choice of grapes and rootstocks. For instance, the Truchard vineyard in Carneros has a fault that has juxtaposed clay soils against Mayacamas volcanic material, and the grower believes the former suits Pinot Noir, Chardonnay and Merlot, while the latter are preferred by Syrah, Zinfandel and Roussanne.

Some Alsatians make a point of matching cultivars with the soil, for example Pinot Noir vines on more calcareous materials and Auxerrois on heavier, clayey soils derived from volcanic rocks. Two roughly parallel faults define the chief vineyard area of Alsace, with subsidiary faults splaying and weaving in between. Hence, together with a series of criss-crossing other faults, the region's geology is divided into an unusually intricate patchwork. Consequently, in practice, individual cultivars are grown on all kinds of ground; Grand Cru Riesling vineyards, for example, are found on a wide range of igneous, metamorphic and sedimentary bedrock.

The soil differences across a fault undoubtedly have a role in the vines' growth, but many commentators believe that they also affect the character of the finished wines. Thus, according to one commentator, Alsatian Riesling wines from calcareous soils have power and complexity, and in some examples fruit and finesse, but where faulting has brought schist and granite to the vineyard the wines have a 'discrete elegance and virility'. In the Asnella vineyard in Portugal's Vinho Verde DOC, a fault has brought together two very different soil types, schist and granite. According to the producer, 'the granite lends richness and weight, while the schist gives minerality, lift and precision'.

The faults in Nervi's Valferana vineyard in Italy's Gattinara DOCG (a UNESCO-listed Geopark) are said to give the soils an 'extraordinary mineral content'. I have to say, though, that such claims never explain how this works: it seems to be taken as somehow self-evident that it is the geology that is responsible for the wine differences, as I discussed in Chapter 1.

Shaping the land

Faults can be fundamental to the very form of a vine-growing region. For the rocks either side of the fault can have contrasting durabilities – different resistances to weathering and erosion – and this can have a profound effect on the landscape. One situation is where the more resistant rocks make higher land and vice versa, with the slopes separating the two, a fault escarpment on the regional scale, providing prime sites for vine growing.

The great vinous example is Burgundy's Saône fault zone. Here, the land to the west of the fault was uplifted to give the high plateaus of the Hautes Côtes whereas the down-dropped plains to the east grow vines traditionally producing Bourgogne ordinaire (now Coteaux Bourguignons). But it is the relatively narrow, roughly east-facing zone separating these two blocks, the fault escarpment, that is the vinous honeypot known as the Côte d'Or.

There are numerous other examples around the world. The stepped slopes of vineyard sites around Woodside and Lobethal, in Australia's Adelaide Hills wine region, result from a series of faults uplifting the Mount Lofty Ranges to the east from the coastal plains to the west. The faults continue southwards, to form the eastern boundary of McLaren Vale. The Willunga Fault dominates here, and occasionally produces minor earth tremors.

Vineyards are dotted along the Darling Scarp in Western Australia, which forms a distinct dividing line between the Perth Hills wine region on top of the escarpment, and the Swan District to the west below. The escarpment is primarily caused by the major Darling Fault system, which extends for at least 1,500 kilometres (930 miles) from the south coast near Pemberton north to Shark Bay (famous for its stromatolites – living equivalents of fossils over 3.5 billion years old). The Darling Fault itself is several hundred million years old, and through time the escarpment has been eroded east by about 15 kilometres (nine miles) from its initial position.

Another situation is where vineyards are located in valleys that are formed at least partly as a result of faulting. In some cases, the valley is the result of erosion following the fault zone itself, picking on rocks weakened by breakage. One example of this is the Breede River Valley, South Africa's most prolific producing region. Here, summers are arid and hot, hence much of the wine is used for brandy. However, fine wines are made towards the southeast, where cooler, moist air can channel up the valley from the Indian Ocean.

The reason for this valley being here? It's the result of erosion by the Breede River along the Worcester Fault system, which extends southeast from the Watervalberge west of Wolseley, through Worcester and Robertson to Swellendam and beyond. It first formed nearly 200 million years ago but has moved the rocks intermittently ever since. Vineyard workers felt tremors at Ceres in 1969, and again near Worcester in 2010.

In other cases, faults have brought a strip of softer rocks next to harder rocks that now make upland areas bordering a valley. The eastern flanks of the Hood River Valley, Oregon, in the Columbia Gorge AVA, are defined by the north–south Hood River Fault, which has displaced the upstanding rocks to the east by over 600 metres (1,970 feet). Vineyards near to the escarpment gain some protection from the famous winds funnelling down the Columbia River Gorge, one of the windiest places in the northwest USA and a mecca for windsurfers.

The northeastern border of Australia's Barossa Valley is defined by the Stockwell Fault. Starting around 65 million years ago, tough, old rocks were uplifted to the east of the fault to make the Barossa Ranges while subsidence on the western side allowed sediments to accumulate. These yielded soils with greater water-holding capacity and potential rooting depth than those to the east of the fault, no small reason why in the hot, dry climate of the valley, ungrafted, unirrigated vines have been able to survive, in some cases for well over a century.

The west coast of New Zealand's South Island parallels a major fracture called the Alpine Fault, which towards the north splays into a number of sub-faults. These have a lot to do with defining the landscape of the surrounding wine districts. One strand defines the eastern extent of the Nelson wine region; in Marlborough, one fork has given rise to the Awatere Valley, and a more northerly splay – the Wairau Fault – runs through Renwick and Blenheim to form the Wairau Valley. The Wairau River, engorged with clay from mountain glaciers, follows this Wairau Fault down to the coast, where the sea has eroded the weakened rocks to form a marked bay. When Captain James Cook moored there in 1770, he decided to name it after the muddy waters brought there by the river. He called it Cloudy Bay.

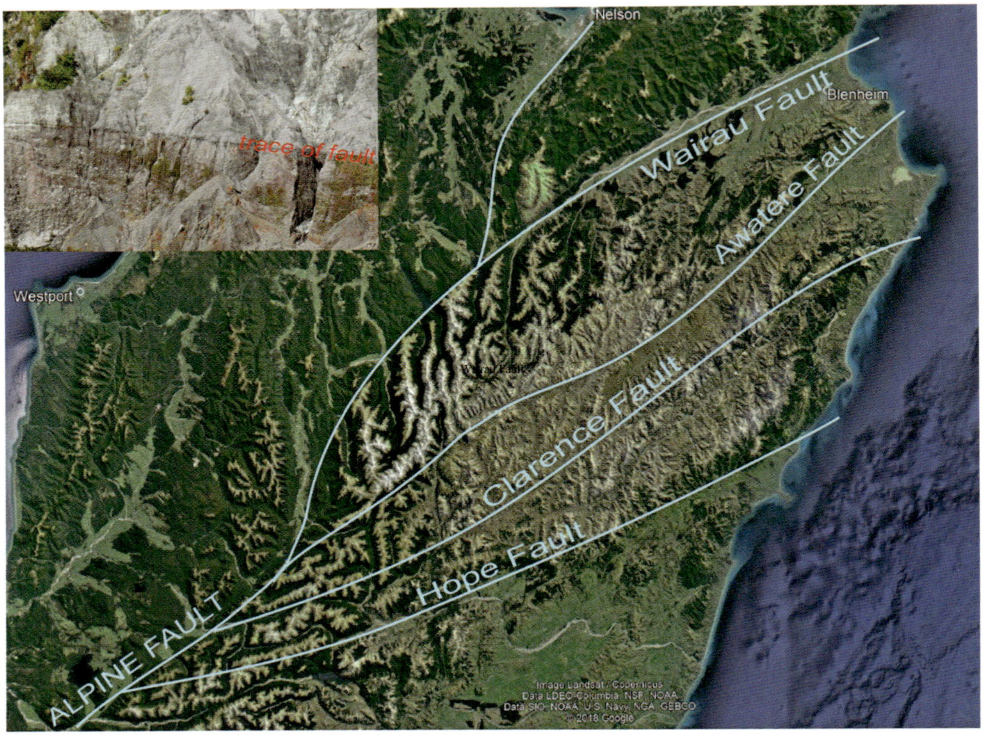

Why do we get earthquakes?

In recent years, earthquakes have taken on a new relevance to viticulture, with tremors that have damaged vineyards and winery installations in several parts of the world. The phenomenon has, of course, been known since ancient times, though understanding that earthquakes are related to movement along geological faults proved very elusive. In fact, despite centuries of pondering in the Old World it fell to the United States – including wine-producing areas of California – to make some crucial contributions.

For millennia, many people assumed that earthquakes were marking the wrath of a god – divine anger over whatever happened to be seen as the heresy of the moment. Other folk followed Aristotle, for whom earthquakes meant that a living, flatulent Earth was breaking 'intestinal wind'.

ABOVE: Satellite image of New Zealand's Alpine Fault showing how its splays determine wine regions such as Nelson, Marlborough and Awatere. The inset shows a rare exposure of this fault, in Gaunt Creek, Whataroa, here horizontally separating white from pale brown rocks.

Naturally, modern geologists had to reject such irrational causes, but they were stuck for a convincing alternative. To many early scientists it seemed, because earthquakes affect a wide area, that the atmosphere had to be involved somehow. People talked of 'earthquake weather', though it was unclear what this was. In literature, Winnie-the-Pooh saw Tigger hiding 'because Eeyore said that nice weather like this is earthquake weather!... Hundred-Acre Wood might have an earthquake today! I'm staying here until it's over!' One or two early scientists did come tantalizingly close, though, to making a fault–earthquake connection. In 1760, for instance, the Englishman John Michell argued that earthquakes must travel through the solid earth as physical waves, and he included in his discussion a diagram of faulted rocks – while making no explicit link.

Knowledge improved when the Scotsman James Forbes (after whom the Aiguille Forbes, above Chamonix, and the Forbes Glacier and Forbes River in New Zealand are named) demonstrated in 1841 a mechanical device for measuring earthquake waves. The instrument was called (by another worker, John Milne) a 'seismometer'. It was the first usage of a 'seismo-' word in English. Derived from the Greek word for shaking, this prefix is now standard for many terms to do with earthquakes, and it's a further instance (together with, for example, 'aftershock' and 'epicentre') of fault/earthquake words that have entered the public domain. For instance: 'In a seismic shift in US policy, President Trump…'

In the later 19th century the spotlight swung to the USA. One night in March 1872, John Muir was sleeping in his cabin in the Yosemite Valley and awoke to a series of 'shocks so violent and varied' that he ran outside, 'both glad and frightened, shouting, A noble earthquake!'. He had experienced what is now called the Owens Valley earthquake, one of the largest in US history. Muir noted later that: 'After the ground began to calm, I ran across the meadow to the river to see in what direction it was flowing and was glad to find that *down* the valley was still down.'

Grove Karl Gilbert, the first senior geologist of the fledgling US Geological Survey, noted that fault escarpments had been produced during this Owens Valley earthquake and began to regard these little faults as integral to the earthquake and not just a side effect. Similar ideas were developing in Japan, but without much international recognition; in the West it was the 1906 San Francisco earthquake that finally nailed it.

Following that event, Gilbert visited Marin County, just north of San Francisco Bay, and in places such as Paradise Valley carefully documented a range of newly formed features. Crucially, he argued compellingly that all these effects and the great earthquake were intimately related. Then another member of the official team tasked with analysing the earthquake, Harry Reid, demonstrated mathematically that the earthquake was the result of stresses accumulating within the ground over many years and then suddenly rupturing along the faults, abruptly discharging the stored-up energy. Thus Paradise Valley, just west of the North Coast AVA and now a key producer of organic food for San Francisco's restaurants, was pivotal in establishing our modern understanding of earthquakes.

Thirty years later the US contribution was extended further, when Charles Richter devised in California a scheme for comparing the magnitude of earthquakes. And from it, seismology was to provide yet another technical term that has become familiar: as in 'buying Clos des Lambrays doesn't really register on the wine investment Richter scale'; the 'Jacques Selosse Brut Rosé is off the Richter scale for rarity, depth and supreme enjoyment'; and 'the reverberative finishing intensity of Raveneau's Les Clos needs to be measured on the Richter scale'.

The final official report on the 1906 earthquake demonstrated that the San Andreas Fault was still active – an astonishing 477 kilometres (300 miles) of the fault had ruptured to trigger the earthquake. Moreover, it showed that the fault stretches all the way south to the Mexico border and beyond. So this was indeed a major structure in the Earth's crust! And curiously, virtually all the numerous horizontal offsets, when viewed by looking across the fault to the other side, are a move to the right. But this fits exactly with our modern understanding: the San Andreas Fault is the junction of two tectonic plates, with the region to northeast being part of the southeast-moving North American Plate and, slipping past it towards the northwest, is the Pacific Plate. Each little crack is replicating in miniature the movement of these shifting tectonic plates.

Earthquakes and viticulture

An analogous situation to the San Andreas Fault occurs with New Zealand's Alpine Fault, which we now know is caused by the Australian Plate moving northeast, to the right looking across the fault, past the Pacific Plate to the southeast. The fault is active, with the rocks being displaced at rates anywhere between three and over 30 millimetres a year. Every now and then this leads to a major earthquake. Past events are recorded in displaced alluvium and they seem to occur very roughly every 330 years. The most recent one was in 1717…

Sub-faults form the Marlborough Fault system and some of these shifted in November 2016, giving the second-largest earthquake ever recorded in New Zealand. Considerable damage was done to vineyards and wineries; more than a thousand storage tanks were damaged, and one estimate had wine being spilled that could have filled more than five million bottles.

One of the world's most seismically active countries is Chile, its very shape being defined by faults splaying off the north–south junction between the Nazca and South American tectonic plates. Here, however, the plates aren't rubbing past each other sideways, as in the San Andreas and much of the Alpine Fault, but the oceanic Nazca Plate is driving at a shallow angle down beneath the continental plate that is thrusting over it. It's this arrangement, common around much of the Pacific Ocean, that produces the world's biggest earthquakes: in Chile eight major events took place during the 20th century alone, including the largest earthquake ever recorded anywhere, the 1960 event centred on Lumaco, in the Malleco wine region.

The fault rupture that generated that earthquake was around 1,000 kilometres (620 miles) long, almost equivalent to the entire length of the San Andreas Fault! Major faults splay upwards off this plate junction, such as the San Ramón Fault, which runs north–south to the east of Santiago. It separates, for example, the Maipo and Cachapoal wine regions from the Andes to the east. The 2010 event that caused so much damage to wineries and vineyards was centred on the inter-plate fault itself, just offshore from the Maule wine district.

The North Anatolian Fault system runs for more than 1,200 kilometres (750 miles) across Turkey, more or less east–west, through the Diren vineyards north of Tokat in the east, westwards through eastern Marmara and the vineyards south of Sakarya. It helps demarcate the great inlet of the Marmara Sea, with Istanbul and Thrace to the north. Very worryingly, in recent centuries earthquake activity along the fault seems to have been moving progressively west.

The most recent major event, the 1969 Gulf of Ismit earthquake and associated tsunami – yes, tsunamis are not restricted to open oceans – killed tens of thousands of people and some seismologists reckon the next episode is imminent and will be centred yet further west – right in the Sea of Marmara. Such an event would threaten both nearby Istanbul, the world's fifth most populous city, and the Thrace region immediately to the north, the area of Turkey with the greatest concentration of wineries.

In the latter part of the 20th century, seismologists put much effort into trying to find a way of predicting earthquakes. Results proved elusive, at least in constraining the timing sufficiently to be practically useful, such as for advising the evacuation of an area. So the spotlight has swung to earthquake 'preparedness', such as engineering structures to resist the tremors and by educating the populace in how best to respond. In the light of earthquake damage in recent years in such important viticultural areas as Napa, Marlborough and central Chile, this thinking is being reflected in the wine world.

Winery buildings are now being 'seismically retrofitted', such as by stiffening them with steel frames but with flexible linkage to a robust base pad. Storage systems, a principal source of winery losses in the recent events, are being rethought. For example, the traditional stacking of barrels in free-standing pairs up to six levels high is being replaced by the broader base offered by four or even six barrels side-by-side, held in interlocking rigid frameworks strapped to walls and ceilings.

Some designs have the framework sitting on ball-bearings in order to ride with the earthquake rather than topple over. A lot of attention has been given to the design and construction of storage tanks; after all, a rocking tank with 40,000 litres of wine sloshing around in it is potentially a very dangerous weapon! Cellar workers are instructed in procedures such as immediately shutting off power and water, and sheltering beneath some robust support within the building. There has even been talk of installing 'shark cages', heavy-duty steel cages where workers can take refuge if bottles and barrels start flying around.

The fascinating San Andreas Fault

The fishing village of Anstruther on Scotland's east coast is well used to loss of life in North Sea storms, together with the expectation that sons will follow in a family's tradition of going to sea no matter what. So when young Andrew Lawson's father was lost to shipwreck, his mother decided that, rather than have her son exposed to such risks, she would leave Anstruther altogether and emigrate (in 1866).

Little did she know that Andrew was to become one of the University of California's first geology professors and, among a host of other things, would discover a geological feature that he called the San Andreas Fault. (The name, Professor Lawson explained, came from the little lake called Laguna San Andrés that lay right on the fault, though some contemporaries, noting his growing ego and his first name, thought he had another reason.)

The fault, surely now one of the best known of all geological phenomena, runs southwards from the Pacific Coast at Point Arena, west of Anderson Valley and Ukiah. It approximately coincides with the western border of the Mendocino Ridge AVA, and southwards near San Juan Batista it makes the northeast boundary of the Monterey AVA. It has numerous splays along its length, but especially so in the San Francisco Bay area and to the north, around Sonoma and Napa. One example is the West Napa Fault, which more or less defines the western edge of the Napa Valley: it was renewed slip along this structure that triggered the damaging Napa earthquake of August 2014.

The fault continues for more than 1,000 kilometres (620 miles) southwards to the Salton Trough on the Mexican border, and affects a host of Californian AVAs and vineyards one way and another. Most of its course is characterized by valleys, but in places tougher bedrock gives localized hills.

The foggy coastal hills of the Sonoma Coast AVA, for example, are cut right through by the main fault, very close to wineries such as Hirsch, Seaview and Flowers. Consequently, the Hirsch vineyards, according to the proprietors, have a 'fearsome diversity of soils, aspects, elevations and microclimates'; the 'degree of fragmentation is unmatched even by the famously subdivided vineyards of Burgundy's Côte de Nuits'. The winery's signature Pinot Noir is labelled San Andreas Fault. The Flowers winery claims that the pulverized bedrock gives 'incredible diversity' to its soils, and leads to low-yielding vines and small grape clusters 'with great concentration'.

The main fault also cuts through the Santa Cruz Mountains; the parking area and tasting room at the Savannah-Chanelle winery, built a mere six years after the 1906 earthquake, are located right on the structure. Nearby streamlets still show offsets from the faulting, once again towards the right. Five kilometres (three miles) to the northwest is the noted Ridge Winery where, just to the west and located on the fault itself, is the conservation area of Monte Bello, a name well known to lovers of Ridge wines.

DeRose Vineyards, south of Hollister, is also sited right on the San Andreas Fault. A crack in the winery's main building runs through the warehouse between the fermentation tanks and the carefully stacked ageing

FOLLOWING PAGES: Aerial photograph of the San Andreas Fault crossing the Carrizo Plain, California, USA. The land to the west (*left*) of the fault is moving northwards, overall at around 35 millimetres (one and a half inches) a year.

barrels, and continues to move at a rate of about a centimetre (half an inch) each year. Some of the damage can be seen at https://www.wsj.com/articles/SB125936328750267155. If you watch the video, notice the displacement of the floor tiles and a drainage ditch – once again towards the right!

Further south along the fault sits the little settlement of Parkfield, northeast of Paso Robles. It's also located right on the San Andreas Fault: drivers entering the village (population 18) from the west see a road sign that reads 'Now entering the North American Plate' and cross a kinked bridge that straddles the fault. (On crossing the bridge in the other direction, another road sign says: 'Now entering the Pacific Plate.') Between 1857 and 1966 Parkfield was shaken approximately every 22 years and so an array of sensors was installed in order to closely monitor the expected 1988 event. The earthquake duly came – in 2003,

ABOVE: Fermentation vessels at the Wente Winery near Livermore, California, USA, damaged by the 1980 earthquake.

15 years late! It damaged irrigation pipes in the Parkfield Vineyard, toppled cases of wine, and plainly illustrated our inability to predict earthquakes.

However, apart from the continuing challenge of predicting the timing of earthquakes, these days the San Andreas Fault is reasonably well understood. Indeed, our knowledge of faults and earthquakes in general is pretty sophisticated. Well, it is in science. It seems, though, that some folks prefer that it all remains veiled in mystique. For instance, I read that the 1906 San Francisco event was partly due to its astrology (☉ ♈ 06°05′ ☽ ♎ 22°59′ Asc ♑ 25°07′).

Paul Giamatti's character in the much-loved film *Sideways* (2004) may have known his Pinot Noirs and his Santa Rita wineries but in the film *San Andreas* (2015) his scientist studying earthquake prediction doesn't seem to have known much about earthquake prediction. In the film he makes various implausible pronouncements, based on measuring things unknown to seismology. But then, misconceptions are common: California isn't going to fall into the ocean; earthquakes don't create chasms that swallow things; faults don't radiate mysterious paranormal energies; and so on.

Yet the wine world is not immune: 'The energy of the fault permeates everything; it gives the wine its tension'; 'his Côtes du Rhône vineyard is run cosmologically… rebalancing cosmic and telluric forces like acupuncture, at the active points of an underground fault'; 'the vines benefit from intense energies emanating from the vortex pole south of Sedona down through Oak Tree Canyon along a subterranean geologic fault'. It has been a long road from celestial wrath, intestinal winds, earthquake weather and the like to the scientific understanding we have today about faults and earthquakes, but it seems that not everyone is making the journey.

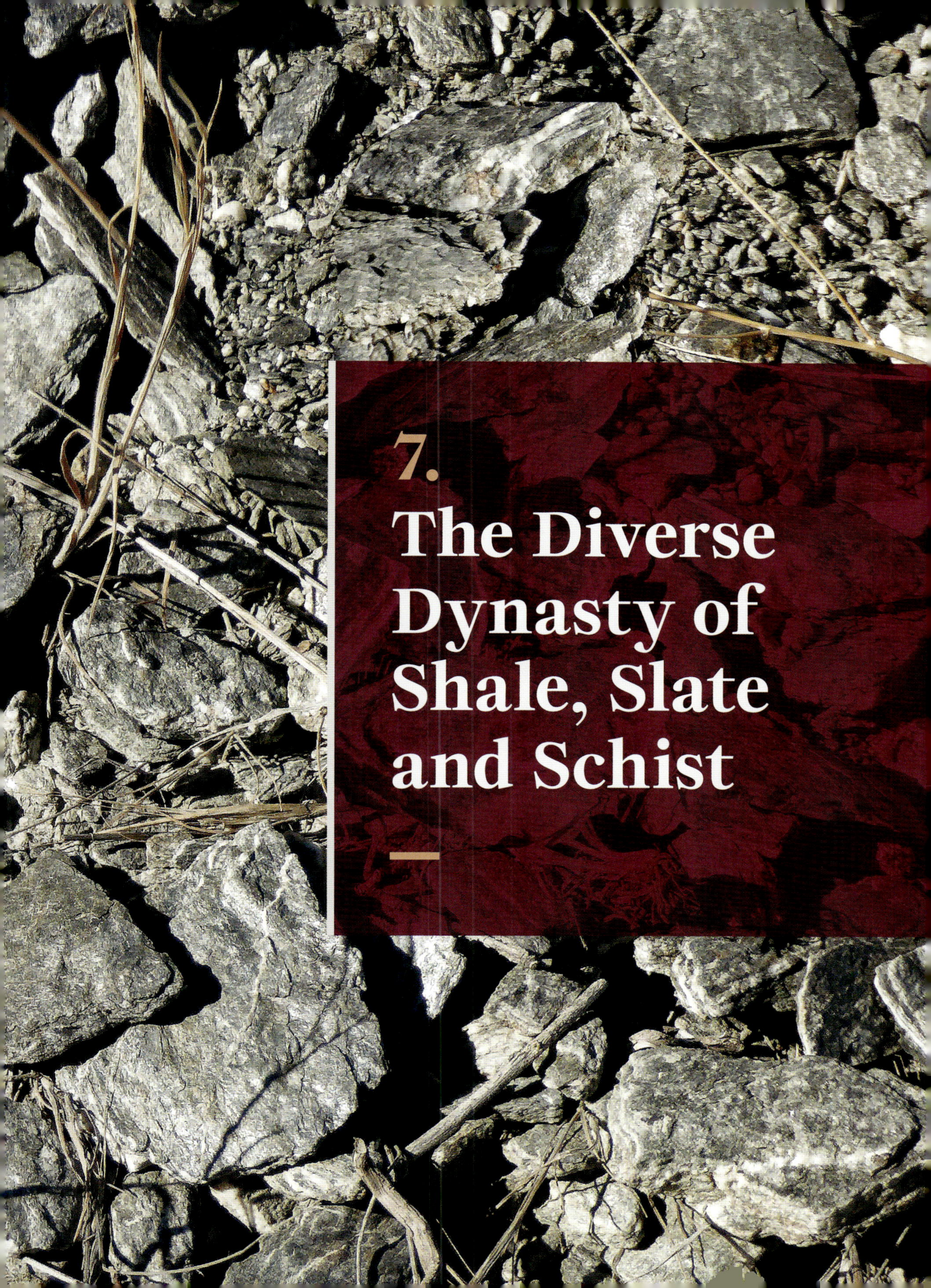

7. The Diverse Dynasty of Shale, Slate and Schist

The Diverse Dynasty of Shale, Slate and Schist

There's a funny thing in the wine world about shale. It's very fashionable these days in wine and vineyard descriptions to remark on the geology – but shale? It's seldom mentioned! How many vineyards can you name that are founded on shale, or producers that boast their wines come from shaley soils? And yet shale is just about the most widespread bedrock on the Earth's land surface and therefore, presumably, must be pretty common in vineyards. It's as though the rock is of little viticultural relevance, or perhaps it just lacks, well, marketing charisma.

What a contrast with slate! This rock, despite having a much more restricted distribution, is eagerly talked about by wine people. In fact, where it does occur in vineyards, notably alongside some of the German rivers, it is treated almost with reverence: many believe it is key to the distinctiveness of the wines produced there. Similarly, schist is not widespread but some growers celebrate and trumpet it for bringing a special quality to their wines. Schist is their touchstone. But shale? Perhaps the problem is its very ordinariness, rather like a sliced white bread of rocks!

So, in the world of wine, we have three distinct rocks with differing auras. In geology, however, things are not like this at all. Geologically speaking, these three rocks are closely related, with shale blurring, in a kind of evolutionary sequence within the Earth, into slate, and slate in turn graduating into schist. In this chapter I will outline how this happens.

I wish also to bring out here another aspect of these three rock names. Each one, though a single word, in fact covers a considerable diversity of materials, spanning various chemical and physical properties and giving rise to wide assortments of vineyard soils. And we will see that while in wine marketing this seems irrelevant for

PREVIOUS PAGES: Close-up of the soil at Chard Farm, Central Otago, New Zealand, showing schist fragments with typ cal shiny lustre.

shale, it leads to promotional strategies for wines from slate that are quite different to those from schist, none of which fits comfortably with the geology of the rocks.

The key to it all: mud

As outlined in Chapter 1, rock at the Earth's surface is continually being weathered. Chemical reactions convert some of the material to fine clay minerals, and at the same time the rock is being physically disintegrated into loose detritus. Inevitably this weathered debris will be moved away, and all the time the grains will be banging into each other, abrading and shattering, and becoming smaller and smaller. It is the resulting fine material – silt and clay – that concerns us here.

This shifting fine sediment may accumulate temporarily in places such as a lake bottom or riverbed (or in a vineyard puddle!) but eventually most of it will reach the coast and ultimately be swirled out to sea, to slowly settle on the sea floor. We are talking here about the silt–clay mixture called *mud*. Hardly glamorous stuff but actually tremendously important for life on Earth and, especially after it has been hardened into rock, a material that impinges on all kinds of human activities.

In fact, the Geological Society in London designated 2015 as 'The Year of Mud'. And it's tremendously widespread: if we include the ocean floor, it is the single most widely distributed geological material at the surface of the solid Earth. Moreover, geologically it is the parent of the three rocks we are going to consider here – shale, slate and schist. So first we need to look a little more closely at how this sloppy mud is transformed into solid rock.

Looking at it very simply, it's like squeezing a lump of wet potter's clay between your hands. At first, it's easy to squeeze out a bit of the moisture. The material then gets that little bit stiffer, because the tiny clay particles are being brought closer together and more efficiently packed, and as you squeeze more, it gets progressively more difficult to squeeze out the remaining water. But if the tightening continues, the clay mass becomes quite tough.

In nature, it's the weight of sediment being added above that does the squeezing, and with all the time available there, chemical bonds start to develop between the increasingly packed particles such that eventually the material becomes hard enough to be called a rock. It normally takes the accumulation of hundreds of metres or more of overlying sediment and the passage of at least thousands of years to fully achieve this in nature and we end up with a material that is properly described as a mudstone. But there's a fascinating accompaniment to all this that will take us to shale.

Shale the mundane

Imagine a leaf falling from a tree. It doesn't dive downwards and land on edge but flutters, with a side-to-side motion, before coming to rest flat on the ground. Additional leaves stack up around it, making a pile with the leaves themselves lying roughly horizontal, parallel to the ground. Any intervening debris may make a thin layer, and overall this will also be arranged parallel to the ground.

The thing is that the clay minerals in mud, though tiny, have a distinctly flaky, leaf-like shape (*see* Chapter 9) so they settle in an analogous way on the seabed. Their parallel alignment will become enhanced as the mud is buried and progressively squeezed from above. There may occasionally be influxes of slightly coarser silt, perhaps owing to a recent rainstorm on land, and this will be deposited in thin layers, across the sea floor.

Thus, in these various ways the body of accumulating muddy sediment takes on an overall planar aspect: horizontal, parallel to the sea bottom. The layers of differing sediment are known by geologists as bedding, and with time the whole mass becomes hardened into a sedimentary rock, characterized by this internal bedding.

If at some later time Earth's internal processes force this buried rock upwards then, coupled with the erosion inevitably taking place at the surface, the rock could be exposed for us to see, even at the floor of a vineyard! And because of its internal structure the clayey parts will have a tendency to break in a direction overall parallel to its bedding – and this is the characteristic of shale. It's a rock hardened from sediment with most of the constituent particles too fine for us to see, and with a tendency to part in rough pieces overall parallel to bedding, the original horizontal of the sea floor. The rock may have undergone some tilting during its uplift but if it's to be called shale, it will roughly follow bedding. And it is this *breakage parallel to bedding* that's the key difference, as we shall see shortly, from slate.

Shale readily fractures into a splintery scree, which can be quite eye-catching. Consequently, the names of topographic features sometimes involve shale, such as Shale Ridge, Shale Peak and Shale Hill, and names such as these have found their way into winery names. Otherwise, the only mention that shale seems to get in wine literature is a casual comment about shaley vineyard soils, usually without much elaboration of why this might matter.

Almost certainly, its chief relevance is to do with drainage, or the lack of it, although the behaviour of disintegrated shale is complicated. On the one hand, a rubble of angular shale fragments will give a very well-drained soil, with all that

Source of some great wines and a million roofs: slate

There is no defined boundary between shale and slate. This is why reports on a particular vineyard area might seem a bit confused if the slates have only reached an early stage of development. It is why some reports on vineyards in the Rheingau, for example, say they have shale soils while other literature declares that they are on slate. Neither is wrong. The Middle Mosel provides a similar example in that it is a classic area of slate quarrying and of famously slatey vineyards, yet the bedrock has not changed greatly because some examples are celebrated for their beautiful and delicately preserved fossils. Normally as slate evolves, any fossils are destroyed so, geologically, at least some of these Mosel rocks are in this intermediate shale–slate 'no-man's land'.

However, as the new minerals continue to form in their new alignment the material takes on a tendency to split in the direction of the new internal structure. This property is termed cleavage, and, because the minerals are tightly aligned but are still small in size, the fracture can be outstandingly even and clean: we now have slate.

Slate is denser and stronger than shale, because of the more efficient packing of its minerals. But the defining difference is its property of cleaving into planar sheets, which, in contrast to shale, can be in any direction and have no relation to bedding. At the same time, the minerals responsible for the cleavage are too fine for us to see. This fineness

LEFT: An example of the slate-based vineyards and slate-roofed buildings of the Mosel: Josephshöfer, Graach an der Mosel, Germany.

means that without a microscope we can't discern why the slate is splitting, and this, as we shall learn shortly, is the key difference from schist.

Slate is, of course, one of the world's great roofing materials. In parts of Europe, it rapidly replaced turf and thatch when the Industrial Revolution made it possible to extract and distribute slate rapidly and cheaply. The rock is found in a number of wine-producing areas around the world: for example near Angers, just north of the river Loire in France (*see* Chapter 8); around Cebreros in the south of Castilla y León, Spain; the Clare Valley, South Australia; the Aconcagua Valley of central Chile; and, apparently, the Hua Hin Hills vineyards of Thailand. These slates are very varied in nature as, exactly in line with shales and their various parentages, slates can have all sorts of mineral compositions; a number of textbook slates around the world were originally volcanic ash.

Even so, the vast majority of slates were originally oceanic mud and now are some shade of grey in colour – although, depending on their precise chemistry, there is a whole range of subtle hues. Slate colours have found their way into wine names, and in the quarrying trade they are given charming names such as royal purple, sea green and heather red. Pant Du vineyard in North Wales uses as a mulch plum-coloured slate waste from huge quarries just a couple of kilometres away. In Valdeorras, northwest Spain, the spectacularly terraced vineyards above the river Sil have a distinctly slatey soil and nearby the bedrock is quarried for the smooth, rather shiny, dark grey slates that are seen on roofs all over the world.

So, is there something special about slate for wine? The belief is probably nowhere more embedded than in the vineyards alongside German rivers such as the Mosel and Rhine, and certainly any wine taster seeing those strikingly steep slopes and vines growing in what seems no more than slate rubble will be put in mind to expect something distinctive. And the wines are, of course, just that, but the idea that the slate can actually be tasted in the wine, as some claim, has to be metaphorical, as I discussed in Chapter 3.

The scientific evidence is that in cool, northern latitudes such as Germany, the chief physical factors influencing the grapes are the climatic effects arising from differing gradients and aspects of the hillslopes, and the proximity to a river. The main contribution of the slate soil itself, as with other kinds of vineyard geology, is its water properties, here allowing the free drainage so important in a moist climate. At the same time, the interlocking, jagged slate fragments give the soil a cohesion that helps reduce erosion on these remarkably steep slopes.

Also, the dark colours of the slates contribute some warming effect and the smoothly cleaved surfaces offer some light reflectivity, both possibly of significance in climates where the grapes are striving to ripen. The chemistry appears relatively unimportant in that most slates weather to yield the full range of nutrients required by vines (with a grower soon correcting any deficiencies that do arise).

There is, however, the intriguing matter of reported differences in wines produced from vines growing near to each other but on soils of differently coloured slate. For example, high in the empty and spectacular Cederberg Mountains of the Western Cape, South Africa, a difference is perceived between wines from Shiraz vines growing in red slate and from those nearby but in grey slate. Riesling wines from the blue slates of Wehlen, in the Middle Mosel, Germany, are promoted as different from those from the red slates there.

So what could be causing this difference? The thermal properties of the soils presumably differ slightly, but this would hardly seem to be greatly significant, certainly in those parts of the world with an adequately warm climate. In other respects the physical properties of the slates are identical: many a slab of slate shows different colours but is physically homogeneous. And the obvious candidate of the chemical differences that are responsible for the colours doesn't really hold up.

The different grey colours of slates are due to variable amounts of carbon that is spread very finely through the rock. These variations can scarcely affect the vines, though, because the amounts are tiny and, of course, plants get their carbon from the atmosphere through photosynthesis. Other colour variations depend on minute differences in trace elements such as titanium, vanadium and chromium, again unlikely to affect vines as the roots are barely able to take up these non-essential, potentially toxic elements, which in wine are in such minuscule amounts that they are not normally reported.

The chief colour influence, certainly for red hues, is iron. Not so much variations in the amount, which is small, but the extent to which it is in its so-called ferric form: it's this that gives the red colour to many rocks and soils, such as the famous terra rossa (Chapter 12). However, although ferric iron is itself insoluble, rarely do slate soils have malnutrition issues. Clearly vines thrive on red slates, absorbing the small amount of iron they need in its soluble, ferrous form. The presence of the ferric iron pigment seems to have no effect on their growth.

I suspect the explanation probably lies not in subtle geological differences in the soil but in the factors that will be changing along with them, such as those to do with the mesoclimate and microbiology (Chapter 1).

Anyway, to another question, a continuation of the one raised earlier: what happens if the slate isn't uplifted and exhumed, but continues to be buried? Well, the ambient temperatures and burial pressures continue to increase, and they act to enhance the processes of forming new minerals, enabling them to grow to a larger size. If this material does eventually find itself at the Earth's surface the minerals will have become big enough for us to see, and the platey ones will clearly show their alignment. But perhaps surprisingly, with these larger particles this rock will no longer break as easily and cleanly as a slate. In other words, we now have a different kind of rock, one known by the ancient name of schist.

Schist: *'l'appellation nature', 'le grand terroir'*, etc

First, the gradational nature of the boundary between slate and schist. As mentioned above, with schist – unlike slate – we can *see* the aligned minerals, and the rock doesn't split with a clean cleavage. But these are subjective judgements. Different geologists will 'draw the line' in different places, such that for one a vineyard may be sited on slate while another would call the rock a schist. An additional complication is that in other European languages the equivalent terms for these rocks have different ranges of meaning. For example, the German word *schiefer* is commonly used for both slate and schist, as is the French word *schiste*, which is also sometimes extended to include shale. So we have to be careful when reading literature on slates and schists that is in English but derived from another language!

In English, a schist is defined by a more or less parallel alignment of minerals that are themselves clearly visible to the unaided eye. Where the structure comprises aligned micas, the schist can almost gleam as the silvery platelets reflect the sunlight. Just as with slate, various non-platey minerals can also be present, not only persisting grains of quartz but ones formed in the new conditions, such as garnet, which commonly stands out as little pimples and lumps.

The blurry nature of the boundary between slate and schist is well illustrated in the Douro region of Portugal and in Priorat, Spain. For both areas, some accounts report a schist bedrock while others call it slate. Again, neither is wrong. For the Douro vineyards, the important thing is that the planes of weakness in the rock are very steeply inclined, allowing both rainwater percolation in wet times and penetration by vine roots. This contrasts with the other rock of the area, the massive granite that barely allows either effect. Much, though not all, of the Priorat area also happens to have planes of weakness in the bedrock that are steeply inclined or vertical.

Just as with its parent rocks, the word 'schist' covers a range of mineral compositions and physical properties, including colour. Those schists that are rich in mica minerals tend to be finely crumbly whereas quartz-rich examples break into large tough slabs. Small amounts of graphite, from organic carbon in the original rock, give a dark grey or blue-black colour.

At Morgon in Beaujolais, the schist-derived, dull red soil is distinctly crumbly (the locals call it 'rotted rock') and at Cascastel-des-Corbières in the Languedoc, the tiny fragments of schist make a black vineyard soil that trickles between your fingers. In northern Corsica, by contrast, some of the schists are much richer in quartz, and the rock breaks into substantial pieces, some of which have been carried away from the fields to build houses. Much of the Dão DOC in Portugal is a plateau of crumbly pink granite soils but in the southwest of the region the bedrock changes, giving rise to absurdly picturesque villages built largely of sheets of a silvery but rusty-weathering schist.

In fact, although schist may seem a somewhat mysterious rock and its name curious (it may have the same origin as the word 'schism'), it has been known and utilized for one thing and another since ancient times. For example, an Egyptian statue in the Louvre, Paris, dating from almost 4,000 years ago, is carved, at least according to the museum catalogue, from schist. Damigero, writing in the second century BCE, recommended mixing *ischistos* with 'women's milk' to counter 'phantasms and hallucinations', and various other inscrutable complaints. And in 620, Isidore of Seville was remarking on how easily the rock called *schistos* breaks, and how it gleams in the light, just like, according to him, saffron.

Schist tends to occur in restricted zones scattered across the wine world. Examples include, in addition to those already mentioned: the Minho and Alentejo regions of Portugal; Sonoma, California; Empedrado in the Maule region of Chile; Swartland, South Africa, and the Barossa Valley, Australia. In view of the disparity of such regions, coupled with the variability of the schist itself, it would seem that the current fashion for reporting that a wine is from 'grapes grown on schist' cannot itself convey much about the wine. In France alone, for instance, schist occurs in such dissimilar places as the Côte-Rôtie, Savennières, Maury, Collioure and Kastelberg in Alsace, and even within just one of these areas the schist can be very diverse.

This idea of a commonality in wines produced from schist makes a big contrast with slate where, as we have seen, the belief is that a small variation in something like colour causes a marked difference in the wines. For example, publicity for Faugères in the Languedoc (from which the quotations heading this section come) says that the personality of its wines is due to the schist bedrock, even though the rock is highly variable and yields soils comprised of anything from crumbly shards, through slabs that can be used for building, to some far too acid for viticulture. Local growers even have various words for these different materials to reflect their differing characteristics for farming. (So I, myself, find it hard to see how a distinctive trait to Faugères wines can be due, primarily at least, to such differing rocks and soils, but that is the assertion.)

There is an organized consortium of wine producers that maintains that all its wines are distinctive because its vineyards are located on schist. So far this initiative includes, in addition to all the French places mentioned above, Saint-Chinian, Banyuls, Coteaux du Cap in the far north of Corsica, and Valais in Switzerland. Notwithstanding the disparity in physical conditions between these regions, and the markedly different cultivars, styles of wine, viticultural and winemaking methods, etc, a study for the consortium *L'Association Terroirs de Schistes* concluded that schist does give its members' wines a discernible character. What is this distinguishing hallmark? The report says that it's 'a nice balance of freshness, nice acidity, supple and silky tannins', with 'minerality and salinity being the most resilient feature'. Hmmm.

The variability of schist is also illustrated by the viticultural region with which in recent years it has perhaps become most closely associated: Central Otago, towards the south of South Island, New Zealand. Much of the material here is silvery grey, but it varies from being quartz-rich and strong, and easily extracted in slabs for building purposes, to crumbly, delicate schists rich in mica. And while most of these were originally muddy sedimentary rocks, there are schists in the area that were originally volcanic ash and are distinctly green in colour because of their chlorite content. So once again, just as with the precursor shale and slate, although Central Otago wines are usually reported simply as 'grown on schist', the phrase covers a multiplicity of rock compositions and properties.

So that is how schist comes about, but it's in no way the end of this saga of rock evolution. The Earth's processes grind on, so it is perfectly legitimate to ask again: what happens if the schist is not exhumed but is buried yet further? Our schists will have formed at temperatures around 300 to 500°C (570–930°F), at

something like 15 to 20 kilometres (9 to 12 miles) below the ground, which to us sounds formidable. But from the planet's viewpoint we're still near the surface, and well within the realm of processes that bury and exhume rocks. The rock is still way below its melting temperature, and the centre of the Earth is another *6,000* kilometres (3,700 miles) away! So, much more can happen to our schist, and the answer to the question is that further gradational changes will take place, going on to produce the material we call *gneiss*. But that's another story.

ABOVE: Schist crags and bedrock at Home Block vineyard, Chard Farm, Central Otago, New Zealand. Home to characterful Chardonnay and Pinot Noir – but a schist character?

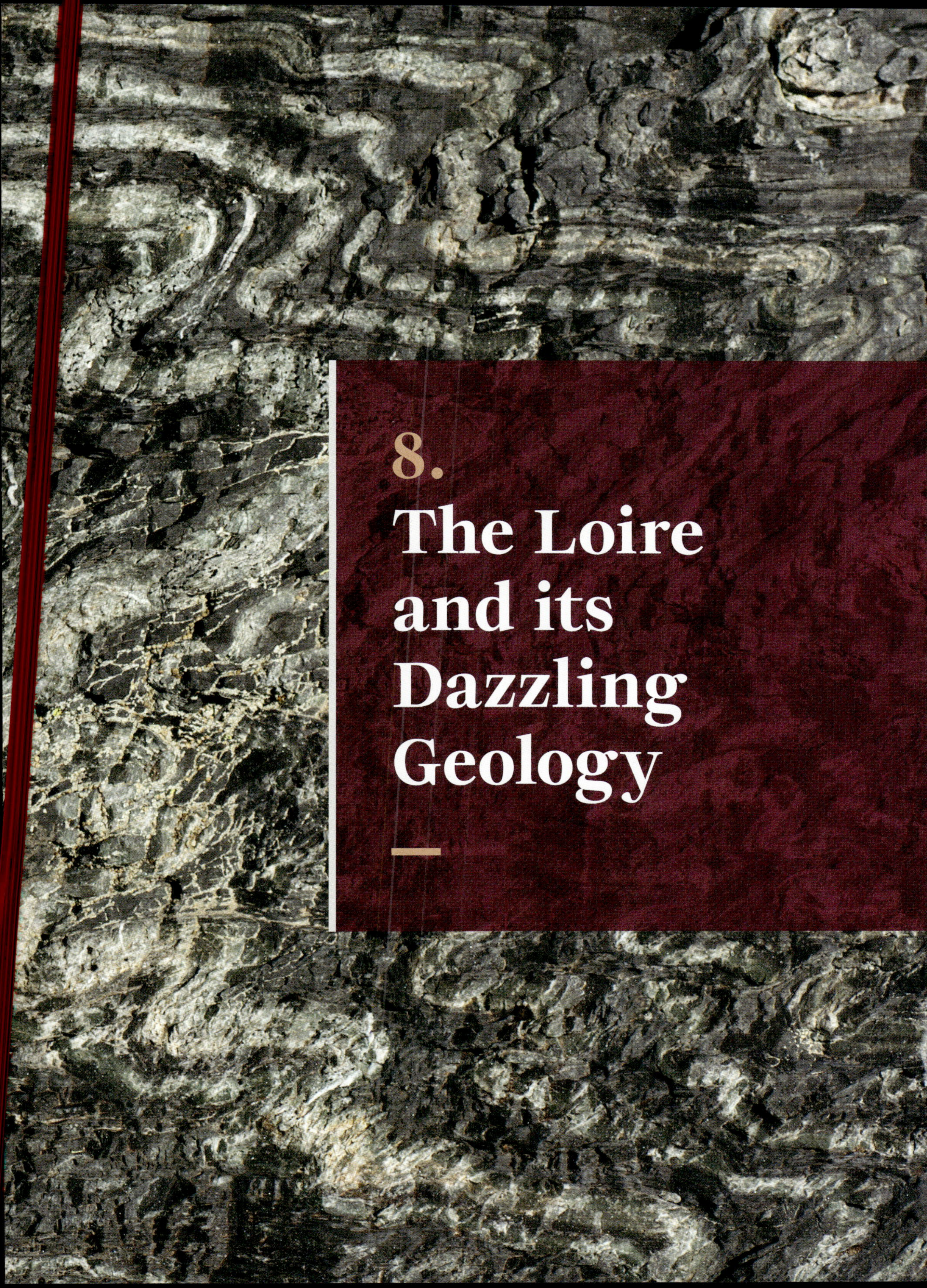

8. The Loire and its Dazzling Geology

The Loire and its Dazzling Geology

Golden eagles and vultures soar overhead, while roaming down in the forests are wild boar, wolves and, by some accounts, lynx. This is the Cévennes, one of France's wildest and least populated regions, and with the darkest of night skies. Fitting, then, that from here flows France's wildest major river, the Loire, barely navigable for parts of its thousand-kilometre course. But ironic that way downstream it acquires the sobriquet 'royal' and glides past celebrated sumptuous châteaux.

It's a contrast further reflected by the wonderfully varied geology of the vineyards along the river's length and the astonishing diversity of their wines. Here, I will outline the geology along the journey, with pauses for a little more detail on some of the vineyard areas – geological vignettes, so to speak, of a wild yet regal river.

A river is born

Volcanic hills, geologically young, dot the Cévennes landscape. One of them, Gerbiers de Jonc, is unusual both in its rock composition (technically, it's a phonolite) and in its being crisscrossed by cracks and narrow fissures. These allow the winter's rainwater to percolate down until they meet old granite-like rocks below, part of what geologists loosely call the 'crystalline basement'. The waters can descend no further and so seep sideways until they eventually emerge at the foot of the mountain as a series of springs.

It's the coalescence of these springs that forms the nascent Loire. Thus, these crystalline basement rocks are instrumental in spawning the river, and as it happens the same rocks floor the river in its final stages, as it meets the Atlantic.

Curiously, the infant Loire leaves Mont Gerbier southwards, flowing towards the Mediterranean a mere 137 kilometres away (85 miles), but after 12 kilometres (seven miles) or so the river swings abruptly north and starts heading for the North Atlantic. It's the result of forces within the Earth that rippled out from the rising Alps and uplifted land to form the Massif Central. They caused land surfaces in the Cévennes to tilt, which resulted in several rivers readjusting their courses. In fact, the Loire has changed its course several times in prehistory, at one time joining the Seine to reach the English Channel.

These same land adjustments led rivers to cut down deeply, generating the spectacular gorges we now see in the Cévennes region. They also gave an overall northward tilting that has led to most of the major tributaries of the Loire – the Allier, Cher, Indre, etc – joining from the south. The land continues to readjust even today: entire portions of the Loire Valley are rising while others are sinking, albeit at rates less than an imperceptible millimetre a year. (Trivial though that sounds, if that movement were to continue uninterrupted for a million years, the land surface would have changed by a kilometre!)

Côtes du Forez

The first wine region of note on the Loire's journey is somewhat enigmatic, little known and short on identity. It's mainly on the slopes of the Côtes du Forez, 15 kilometres (9 miles) or so west across the plain from the river itself. Although an AOC region it's often overlooked; a number of the books I have on French wines don't even give it a mention. Worse than that, just three decades ago, one respected critic was calling Côtes du Forez wines 'rubbery and tasting of boiled sweets', 'inky and farmyardy'. Great improvements have been made since then but the region is still searching for an image.

The soils are very largely derived from the granitic bedrock that underlies the forested hills and so, together with the region's traditional Gamay grape and often preferred practice of carbonic maceration, it was traditionally positioned as something of an alternative to Beaujolais. This didn't really work though, and through time the area under vines has dwindled. There are some patches of basalt and related rocks in the area, localized effusions of the lava associated with the much grander affairs of the Auvergne volcanoes. This is enough, with the recent fashion for so-called 'volcanic wines', for some producers to form an association to promote themselves as 'Loire Volcanique'.

The granitic bedrock is itself a complex business, part of but not identical to the crystalline basement at the Loire's source. These so-called basement rocks are the geological foundation of all of France, the culmination roughly 300 million years ago of billions of years of Earth processes producing rocks that are strikingly crystalline, with well-formed minerals glinting in the sun. Upon them during the geological time that followed were deposited the kinds of sedimentary rocks we are more familiar with: limestones, sandstones and the rest, with volcanic deposits in places.

Granite in these basement rocks will have solidified from molten rock, itself formed deep in the hot Earth from the melting of pre-existing rocks. Sometimes the melting was incomplete such that if this material eventually found its way to the Earth's surface, we would see a melted and unmelted 'mixture'. Such a rock is properly called, from the Greek for mixture, a migmatite.

It's a common bedrock in the Côtes du Forez, along with granite, and appears on several wine names in the area. It sounds a technical name but probably we've all seen examples, as highly polished slabs of light and dark material commonly making swirling patterns, in the walls of banks and prestigious municipal buildings and the floors of airports, Brussels' Zaventem, for instance.

Despite past condemnations of Côtes du Forez wines, there are current lists describing them as 'incredibly delicious' and 'intoxicating and sumptuous'. And some growers are experimenting with new methodologies and trialling cultivars that may be better suited to these soils. Might Côtes du Forez be about to emerge from its past shadows to join the bright stars of the Loire firmament?

The river grows

Down on the plain below the Côtes du Forez, the Loire becomes much boosted near Nevers as it is joined by the Allier. The newly enlarged river lays down broad spreads of sands and gravels, and as it continues northwards it eventually encounters the sedimentary strata that will underlie its bed for the next several hundred kilometres. The crystalline basement is still there but is now buried a kilometre or more below these younger rocks.

These younger strata came about because around 200 million years ago the part of Earth's crust that was to become the North American continent was separating away from Europe, to create the Atlantic Ocean. Hereabouts in Europe, the land sagged such that the sea was able to move in – warm, sunlit waters rich in aquatic life. In the times that followed, first during what geologists refer to as the Triassic period, then the Jurassic, Cretaceous and Tertiary, the extent of the sea varied and hence the water depths were sometimes deeper

and other times shallow enough for reefs, lagoons and tidal flats to form. Consequently, through time a wide range of strata including shales, sandstones, limestones and marls – some rich in fossils - was deposited on the sea floor.

Then roughly a hundred million years ago, the sea started to recede as the region was gradually uplifted by those same Earth stresses that created the Massif Central and the Cévennes. The strata were gently flexed into a broad saucer shape with the youngest strata in the central part, more or less centred on Paris. Hence the arrangement is referred to as the 'Paris Basin'.

The first of these rocks that the river meets are Middle Jurassic in age (around 150 million years old), at about 12 kilometres (seven miles) south of La Charité. Then, by the time the river reaches La Charité itself, it's crossing Upper Jurassic strata, including, further downstream at Pouilly, rocks of the famous Kimmeridgian and Portlandian subdivisions. By the time the river reaches the Côtes du Gien the strata are yet younger (Cretaceous age: around 100 million years) and further on in the Orléans wine district, the bedrock is largely of

ABOVE: The meandering river Loire passes the vineyards of Sancerre: seen here from the Tour des Fiefs, one of six defensive towers of the château that sits atop the town's famous butte.

Tertiary age, some of it less than 20 million years old. But then the geological ages reverse as the river continues westwards and encounters progressively older strata towards the western edge of the Paris Basin.

Of course, in detail things are much more complicated than this, not least owing to localized geological faults having shifted the strata. A pronounced example is at Sancerre, where to the west of the river the Upper Jurassic bedrock has been displaced downwards between several north–south fractures. But curiously, the down-dropped, originally higher rocks of Cretaceous age are tougher – famously flinty – and have resisted erosion. Thus they now form higher land, such as the prominent butte on which the town of Sancerre sits.

Similar tough rocks cap the hill of Saint-Andelain over on the Pouilly side of the river, giving an island of younger (Cretaceous) bedrock surrounded by older, Jurassic, materials. And, of course, the traditional belief is that the these intricately differing rocks have a profound effect on the resulting wines.

The regional uplifts and tiltings of the land surface again become apparent at Orléans, where the river abandons its northward trend towards the English Channel and swings westwards, approaching the renowned wine district of the Touraine. Here, as one example of the kind of geology hereabouts, I focus on one of the lesser-known districts, the AOC Touraine Azay-le-Rideau, between the south bank of the Loire and its tributary the Indre.

Azay-le-Rideau

In 1883, Monsieur Deschand, the local doctor, decided to plant vines just east of the village of Azay-le-Rideau, on the slopes of the Indre. While digging, he stumbled across a cache of several hundred beautifully cast weapons and tools dating from the Bronze Age. And in a nearby vineyard, in 1906, an ancient sarcophagus was unearthed that contained numerous funerary objects and, touchingly, the skeleton of a child clutching a coin, presumably ready to pay the ferryman to cross the Styx into the underworld.

These are just two glimpses of the long human history of the Azay-le-Rideau area, which includes winemaking. A wine press dating from as long ago as the second century was discovered at Baigneux, to the south of Azay-le-Rideau, and in the church of Saint-Martin in Lignières-de-Touraine, to the northwest, a 13th-century fresco illustrates a vigneron happily harvesting his grapes and treading them in his wooden vat.

Even so, Monsieur Deschand was taking a risk planting vines in 1883. The history of winemaking in the district had distinctly fluctuated in success and 1883 was the very time that the dreaded phylloxera was arriving in Touraine. As it turned out, nearly half of the Azay-le-Rideau vineyard area was to be lost to the disease.

But what of the geology in all this? Well, in some ways it seems of little relevance. One published study ranked the factors that appear to have influenced the Azay vineyards over time, and while such things as changes in the local population, drinking habits and transport availability were all significant, the role of the soils here was deemed 'rather weak'. And the soils, mostly derived from calcareous Paris Basin strata of Late Cretaceous age, are still largely ignored by today's growers. For instance, there are limestone soils that elsewhere might be much esteemed, but which here lack vineyards altogether. Yet there are poorly drained clay soils, *a priori* unfavourable to viticulture, that have vines.

Most of the vineyards are located on the slightly elevated plateau between the Indre and Loire rivers, probably because of its advantageous mesoclimate. The soils there are largely stony clays picturesquely known locally as '*perrouches*' (literally, parakeets). Also up on this plateau area are rocks and soils with another intriguing name: *poudingues*. The English translation is 'puddingstones' and in fact this is what such pebbly material used to be called in England, supposedly after the mottled look of an old-fashioned Christmas plum pudding. It seems that even though in England the rock name became replaced by 'conglomerate', the old name sneaked across the Channel to be absorbed into French.

But in a subtle way the geology is crucial for the wine here, at least for its image. For the vast majority of people, the name Azay-le-Rideau means one thing: the celebrated château. It's why over 95 percent of visitors come here. And as with the other great châteaux of the region, it's constructed from the illustrious Touraine bedrock, the *tuffeau blanc*, so well suited for these kinds of splendid buildings. It's a limestone-like rock, somewhat sandy and very porous, easy to extract and shape, and it weathers to give soils well suited to viticulture. Hence, numerous Touraine wines have '*tuffeau*' in their names and on their labels: the châteaux and the stone of their walls have become inextricably linked with the region's wines (Chapter 5).

Actually, the bedrock at Azay-le-Rideau itself consists of strata overlying the *tuffeau blanc*. It's still referred to as *tuffeau* but it contains a potassium- and iron-rich mineral called glauconite that weathers to give the rock a yellow, ochrous look. Hence *tuffeau jaune*. It lacks the desirable, fabric-like sheen of the *tuffeau blanc* and so is used largely for the lesser buildings of the area.

As mentioned, the great advantage of *tuffeau blanc* for building is its high porosity, which makes it soft and easy to work. In fact, the rock's porosity is something like 50 times more than that of granite. This, however, turns out to be a two-edged sword, as the porosity induces the stone to store moisture and this accelerates its weathering. As a result, a number of the Loire châteaux, including Azay-le-Rideau, are now decaying nastily and intensive research is underway on how best to conserve their walls while preserving their characteristic velvety texture.

The Azay-le-Rideau vineyards were included in the 1939 AOC Touraine appellation but in 1953 the district was allowed to append its own name. Whether or not this has been beneficial is debated. One can only speculate how many of the 300,000+ visitors to the château at Azay-le-Rideau are tempted to buy that district's wines just because of the attached name. Maybe some are, and there are now a few enterprising producers attempting to raise the area's profile, but the fact is that the area under vines continues to shrink. Despite the imposing name it now can use, the Azay-le-Rideau vineyard area is today only about a tenth of what it was when Monsieur Deschand was planting his vines.

ABOVE: Troglodyte houses carved into a *tuffeau* rock face, here converted to modern homes in Turquant, Maine-et-Loire, France.

On to Anjou

A few kilometres downstream from Azay-le-Rideau, the Indre joins the Loire. By now the Loire is broad and distinctly braided by sand bars and spits as it flows languidly westwards, to leave Touraine and enter the Saumurois. The river is flanked by broad swaths of modern alluvium, especially to its north, but the bedrock remains largely Late Cretaceous calcareous rocks with some sandstone. They underlie the vineyards all around.

The wines, of course, cover the gamut from red to white, dry to sweet, still to sparkling. The local sandstone is visible in the immense dolmen of Bagneux near Saumur, one of the largest in Europe, but the most conspicuous rock is the *tuffeau blanc*. It's in the walls of the local houses, bridges, the stately castle of Saumur itself, and in the bluffs along the Loire's south bank, riddled with troglodytic caves hosting mushrooms, tourist accommodation and innumerable bottles of wine.

Then, a few kilometres beyond all this, the Loire passes imperceptibly into the region of Anjou, also with wines of assorted colours and levels of sweetness. Most of the great Renaissance châteaux are now left behind but the river's regal sweep continues, crossing the heartland of an empire that once stretched from the Pyrenees up to northernmost England. At first, the bedrock is still the sedimentary strata of the Paris Basin, still sandstones, limestones and the like. They're all pale coloured and hence the district is sometimes referred to as L'Anjou Blanc.

But the strata are close to the very western edge of the Paris Basin's 'saucer', far from the thick deposits that accumulated in its centre, and so they are becoming progressively thinner. Also, erosion has inexorably been stripping away the layers, such that there comes a point in the river's westward course where they are no longer present. It's at the little village of La Daguenière, about six kilometres (four miles) east of Angers. Here, the crystalline basement, which has been there at depth ever since the river left the Massif Central, is once again at the surface, once again visible. The rocks are highly varied but overall they're somewhat dark in colour: we have entered L'Anjou Noir.

Anjou is centred on the historic city of Angers, just north of the Loire itself on the river Maine, an unusual case of a tributary coming in from the north. The district's bisected geology is reflected in the city having in the past been called *Angers la Blanche*, from the pale *tuffeau* walls of its old buildings, but also as *La Ville Noire* because of all the roofs built from the local black slate – Angers was once one of the world's great slate-producing sites.

LEFT: Roofs of Anjou slate are common in the Loire Valley, here in Amboise, Indre-et-Loire, France.

The slates at Anjou are exceptionally fine and can be formed into unusually small and thin plates. This means that as a roofing material they allow for intricate shapes and patterns. No coincidence, then, that a feature of all the Loire châteaux is their elaborate curving roofs and turrets – of Angers slate. So how sad that in 2014 Angers slate miners were told that their work was no longer economical and that the historic mines would close. It ended a tradition started as long ago as the sixth century. And was something of a blow to France's *patrimoine*, given that roof repairs to celebrated châteaux now had to use slates imported from Galicia in Spain.

The slates at Anjou are just one example of the multifarious rocks that stretch westwards from here as part of the crystalline basement, here termed the Armorican Massif. As we saw in the Massif Central, this 'basement' is an amalgam of ancient rocks and geological processes that came to an end around 300 million years ago. In fact, the Armorican Massif is characterized by this antiquity of bedrock rather than its physical height. Uninterrupted erosion has reduced it to an irregular low plain, mostly no more than 50 metres (165 feet) in altitude. But some of the rocks are exceedingly ancient: at Brittany's Locquirec (Finistère) and Trébeurden (Côtes-d'Armor), are the oldest rocks in France, formed over two billion years ago.

The kinds of rock are dazzlingly diverse. It strikes me that some wine commentators love to use geological terms, and seemingly the more technical the better. So the Armorican Massif provides rich pickings. Once again, to provide an illustration I will focus on one area from this part of the Loire.

Savennières

'Like a child's paint box', wrote one commentator when describing the soils at Domaine du Closel, in the heart of the little village of Savennières on the Loire's north bank. The visual effect might have something to do with the limpid luminosity of the river – that master of light JMW Turner came here in 1826 to try to capture it – but it's the assortment of differently coloured pebbles that catches the eye. The Savennières AOC stretches westwards of Angers, over hills and valleys that are at right-angles to the river. And it's particularly at the foot of some of the slopes that fragments of all sorts of rocks have come together.

Some of the pale, shapeless pieces aren't really rock at all, but soft clods of the wind-blown sand that clothes the hilltops. Other pale materials are sandstones and jagged hard rhyolite, a rock exactly like granite except that it is fine-grained, usually because its parent molten material was chilled at the Earth's surface rather than slowly cooling at depth. Another fine-grained, chilled material here is spilite, a sodium-enhanced variety of the dark-coloured rock basalt. Typically, the quenching and the sodium are the result of basalt lavas having erupted into seawater. (Sodium, incidentally, is not a vine nutrient; in fact it can be problematic for vine growth.) In places, the submarine lava erupted into the soft sediments on the sea floor and the two materials became intimately mixed, producing what is called a hyaloclastite. All these geological terms appear in the promotion of local wines.

Snow-white pebbles are of quartz, the common form of silica (silicon dioxide). But silica can also form with extremely fine crystals, discernible only under a microscope, such that tiny amounts of impurities can have a huge effect on the colour. Iron can give a red colour – the material called jasper – found as fragments at Domaine du Closel but more common at Les Baunâts, across the river. Carbon gives a deep blue-black colour. This is called black chert in English, but some English wine writers and promoters use instead a French word – *phtanite*. Perhaps, as with the flint at Sancerre being referred to as *silex*, using the French name adds a touch of exoticism, and perhaps also a little mystique?

There are also pieces of Angers slate and also schist, a similar material that splits less cleanly because it has undergone further internal changes through having been more deeply buried (*see* Chapter 7). And as it happens, in this region there is an additional difference between the slate and the schist: geological age. The Anjou slate formed around 450 million years ago, in a period of time known as the Ordovician, but the schist is well over 100 million years older than that. It was produced during a time called in this part of the world the Brioverian, as a result of particular earth movements known here as the Cadomian. This is all technical stuff; again, I mention these names only because I've seen them all on bottle labels and in wine names, and mentioned enthusiastically in promotional materials and vineyard commentaries.

The wines from Savennières are very largely made from a single grape – Chenin Blanc – but in a range of styles. That's made possible because, apart from differing methodologies, the terrain is varied: there are hilltops and valleys, and slopes of differing exposure, gradient and proximity to the river, all of which leads to an array of mesoclimates. And they're all very different from the wines from over the river – the luscious, sweet wines of the Coteaux de Layon, for instance – even though the soils there are generally similar.

Scientifically, the main contribution of the soils is providing a cohesive strength to curb slippage down the steeper slopes and a suitable drainage in this distinctly Atlantic-influenced climate. However, there are plenty of claims that the types of soil have special influence for the wines. I read that, for example, 'the schist soils bring a seriousness to the wine', 'quartz and rhyolite play a role in the presence of acidity' and 'sandstone is a factor of powerfulness'. As for phtanite, well, it gives 'amazing *passerillage*' (*sic*) and 'conveys to the wines their mystical complex aromas'. They're titillating passages but add no indication of how the soils do these things.

The mysterious rocks of Muscadet

Salty breezes from the Atlantic become increasingly noticeable as the river sweeps ever westwards, past Ancenis towards Nantes. This is the Muscadet AOC, whose vineyards are of most interest to the south of the Loire and most esteemed in the vicinity of the rivers Sèvre and Maine – yet two more southern tributaries. The bedrock throughout the area is still the crystalline rock of the Armorican Massif, which still presents a treasure house of esoteric rock names. Moreover, these are the kinds of rock that can change over very short distances.

On January 25th 1799 the region was rocked by an earthquake. That's not particularly unusual here – there was a 4.5 magnitude quake as recently as 2020 – but this one was exceptionally strong, causing widespread damage to the buildings along the Loire. It was the biggest ever known hereabouts and remains one of the six most powerful in all of France. I'm mentioning this in order to point out the importance here of major geological faults, the source of these tremors. As with all faults, the rocks either side have moved, hence very different rocks have been brought right next to each other (Chapter 6).

There's an example just a kilometre north of Clisson, where the Sèvre is joined by the little Ruisseau du Chaintreau. One of these major faults passes exactly through here and has brought the granite that defines the Clisson Cru Communal directly next to a rock called gabbro, which then extends northwards through the Gorges and Le Pallet *crus*. Gabbro, the hallmark of the Gorges Cru Communal, is a dense interlocking of coarse crystals resulting from slow cooling and solidification of molten rock down below the Earth's surface, just like granite except that its dark constituent minerals make it almost black.

Serious Muscadet producers have long fought the cheap and cheerful image of their wines. (I read that, staggeringly, the average price paid for a bottle of Muscadet in France is less than €4.) One success has been the better quality resulting from leaving the fermenting must on its lees (fermentation residues) for longer than is usual elsewhere. Consequently, the applications for the higher status of Cru Communal were based on extending lees ageing further, together with restricting yields and specifying the vineyard bedrocks.

There's a great deal of published scientific research on the effects and merits (and demerits) of lees ageing and of yields; however, there isn't for the role that bedrock might have. But plenty of passionate producers believe that it's the geology of their vineyards that gives distinctiveness to their wines.

Most of the *crus communaux* have several different kinds of bedrock and here's where yet more arcane names come in. We've already met schist and migmatite. I think the name gneiss is reasonably well known, as is its originating from marked change through increased temperatures and pressures when buried below the Earth's surface. But the defining characteristic of gneiss is its alternating light- and dark-coloured bands, and understanding just why these developed is a highly technical matter (stress-modulated diffusive mass transfer anyone?). More pertinent is the question of why this is relevant to grapevines, as wine names that include the term imply. Most gneiss is pretty much the same as granite, apart from the banding.

Some producers label their wines as 'orthogneiss'. I suspect many a geologist would struggle with that one. It means a gneiss that was originally a molten rock, though in view of the extreme changes involved in producing a gneiss it needs pretty clever geological sleuthing to determine the original rock. And again, what that distant origin means for vines is unclear. (Once again, I am not disputing in any of this that the wines from different sites are dissimilar but questioning if it's the subtle changes in geology that are responsible.)

A particularly interesting, almost charismatic rock – serpentinite (OK, I'll come clean: it was the subject of my doctoral dissertation and its picture takes pride of place on the title pages of this chapter) – is found in the Goulaine *cru*. Much of the land here is marsh but along route D105 southwest of Le Loroux-Bottereau, at the bridge over the Goulaine River, is a small but prominent, vine-covered hill, Butte de la Roche. It consists of various rocks with names such as leptinite, amphibolite and pegmatite – esoteric perhaps, but all mentioned in wine writings. And along the crest of the hill is the beguiling serpentinite. This waxy, snakeskin-like rock is rich in magnesium and is unusual because it can actually be inimical to vines and some other plants: botanists talk of 'serpentinite barrens'. The vines manage here though, perhaps because they are grafted onto Couderc 3309 rootstock, which copes well with all the magnesium. Growers claim that the viticultural challenges are worth it because the stressed vines produce concentrated fruit with distinctive character.

Grapevines grow yet further seawards, even west of Bourgneuf-en-Retz, southwest of Pornic, literally a stone's throw from the ocean. So the Muscadet produced from these vines is truly a maritime wine. Route D213, the 'Route Bleue', heads north from here along the coast to the grand bridge that crosses the Loire, across to Saint-Nazaire. The bridge was the longest in France at the time of its construction in 1975, and here the river's flow as it approaches the ocean can be truly whopping, by far the greatest in all of France (well, unless you count the Rhine). And of all that swirling Loire water some, just some, will once upon a time have fallen as rain in the Cévennes, on the slopes of Mount Gerbiers-de-Jonc.

OPPOSITE: An example, here from Saint-Malo, Ille-et-Vilaine, France, of migmatite, a rock also found in vineyards of both the upper (Côtes du Forez) and lower (Muscadet) Loire. Field of view 50 cm (20 inches). Migmatites ('mixed rocks') were once partially melted: in this example the paler parts were molten while the darker parts remained solid.

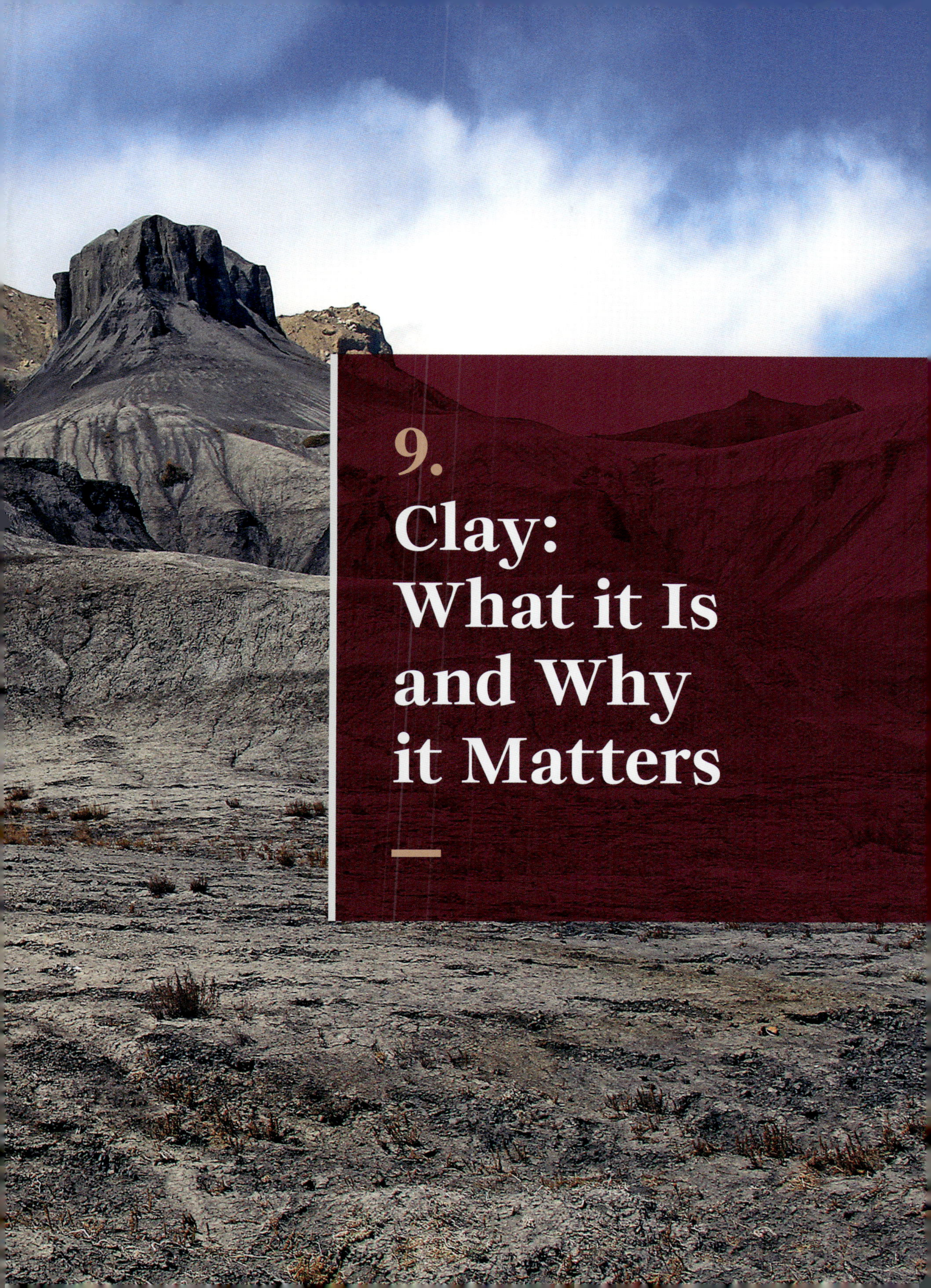

9. Clay: What it Is and Why it Matters

Clay: What it Is and Why it Matters

Clay is familiar stuff. We probably all as children took a lump of clay and modelled it into a little animal or something pot-like. We probably all know that it's baked to make bricks and tiles, jugs and plates. In fact, it's a material used commercially in enormous quantities: in landfill linings, iron foundries, brake linings, drilling muds, animal housing, explosives, and so on, and so forth.

And, of course, its importance for wine can hardly be overstated, with its crucial influence on the nutritional status and drainage properties of vineyard soils. But also, in the wine world, it's used for such things as a sunscreen for vine foliage, for clarifying wine, and for growing vine cuttings. Clay foundations can affect the very stability of winery buildings. Even fermentation vessels, in recent years so fashionably made of oak or stainless steel, are for some producers now shifting back to the material used 8,000 years ago – clay.

Yet despite all this it's a rather mysterious substance. What is it exactly? What is it made of? Where does it come from? And why does it behave in such unique ways? In this chapter, I will discuss some aspects of this distinctive substance that are relevant to wine, and suggest that understanding just two fundamental attributes of clay goes a very long way to explaining why it is so special.

Intricate architecture on a tiny scale

In to order understand clay, we have to peer down to the very atoms that make it. It's made primarily of silicon, oxygen, hydrogen and aluminium but here's the

PREVIOUS PAGES: Bentonite Hills, Utah, USA. The hills are mainly composed of shale but contain layers of volcanic ash that has broken down to form the clay mineral bentonite.

first important thing: the elements are arranged in sheets. That sounds innocuous enough, but it really is one of the keys to why clay is so important. Sheets of inter-joined silicon and oxygen are stacked on top of each other, and held together by the other elements in between to make a solid compound. Reflecting this sheet arrangement, each particle has an overall platey shape; an individual clay mineral is often pictured as a flake.

We can think of the sheets of atoms in a clay mineral as a multi-decked sandwich, with the silicon and oxygen as the slices of bread and the other elements providing the fillings. And in the same way that sandwich fillings vary in how well they hold the bread slices together, so the silicon–oxygen sheets in clay vary in their inter-linking, some being weakly held together and others firmly bonded. It's that simple, but I will come shortly to its crucial repercussions.

And the other fundamental feature of clay? Each flake is tiny, very tiny. An individual flake of clay mineral is at the most 0.002 millimetres (two thousandths of a millimetre) in length; most are even smaller. That's roughly a hundredth of the thickness of a human hair, so obviously we can't see the actual clay flakes – that requires advanced laboratory instruments.

Straight away we can see consequences: such tiny particles will easily clog the pores and conduits in soil, shutting down the drainage. And it's why that lump of modelling clay is so malleable: it comprises innumerable invisible flakes that easily slip past each other so that the whole mass is easily moulded.

It's also why the very finest rock debris, with particles smaller than those in silt and much smaller than sand, is simply referred to as 'clay'. It's the main constituent of mud. So we can think of clay as loose sediment, a soil, a lump, or even a weak rock, but those substances are all aggregates of countless, indiscernible tiny flakes of clay minerals.

Taking the sandwich analogy just one more step, a sandwich shop might arrange its products according to the nature of the bread – say wholemeal, white, brown and so on – and then by the different fillings. There could be quite a range, but, of course, they're all sandwiches. In the same way, mineralogists recognize different kinds of clay according to the nature of the sheets, the way they are stacked up, and the kinds and positions of the elements in between. This leads to a daunting number of different kinds of clay, different clay minerals, each with its own name and particular properties. Here I will single out just a few examples from this extensive range for a closer look. First up, kaolin. It's perhaps the most familiar type of clay, as it appears on the labels of everyday things such as pharmaceuticals (especially, these days, in alternative medicines), cosmetics and garden products.

A secret from the mystic East

Jean-Baptiste Darnet was a leading 18th-century surgeon in Saint-Yrieix, Haute-Vienne, France, so he always had to appear well turned-out. He therefore appreciated the particularly crisp, clean laundry his wife provided him with. And one day he happened to learn her laundering secret: she always added to the wash some of the white clay from the ground around their property. Puzzled, the surgeon sent a sample of this intriguing clay to an apothecary friend for analysis, eventually to learn that he had stumbled across something very special, a kind of holy grail of the time. It was kaolin.

Ever since porcelain figurines began to be brought back from the East by early explorers like Marco Polo, Europeans had sought to replicate the fine product made in China. That porcelain had a purity of colour, translucency, strength and hardness far superior to the soft pottery being made in Europe at the time, at places such as Sèvres, near Paris. However, the materials and methods used in China were jealously guarded; they were a mystery in Europe.

Eventually, however, in 1712, the Jesuit missionary François d'Entrecolles managed to spirit away some samples of the secret raw substance and get them back to France. There they were eagerly analysed and the results published, which immediately prompted a hunt for a Western source of similar material.

In England, Josiah Wedgwood became interested, and even attempted to import and test the fabled 'white burning clay' of Cherokee tribes living deep in the wilds of the southern Appalachians. But Monsieur Darnet had found a supply right in his garden in France. Other sources were soon discovered, and a new European industry was born. Before the end of the century, the clay at Saint-Yrieix was being sent to Sèvres, and was supplying the factories that were starting up at nearby Limoges. Around the same time, similar clay was found at places like Meissen in Germany and Cornwall in England, much to the delight of Mr Wedgwood, and of Mr Spode, Mr Beswick, Mr Doulton and the rest.

But all these clayey materials retained the name of the place in China from which Père d'Entrecolles had procured his original samples, the village of Kaulin, about 45 kilometres (28 miles) northeast of Jingdezhen, in Jiangxi province. An alternative transcription of the name is Kaolin and this version was to become the widely adopted name for the clay. The European porcelain made from the clay became universally known simply as 'china'.

Kaolin is so suited to this use partly because it consists of – as clay minerals go – relatively large, rigid flakes with a relatively simple chemistry. The resulting bonding is strong, with none of the complications that arise in other clays that involve various metals. The bonds are strengthened further on firing, to produce the thin, strong, translucent sheets characteristic of fine china. In practice, there are usually various impurities present in kaolin, and certainly there are in so-called kaolin vineyard soils; geologists refer to the pure mineral as *kaolinite*.

We can use kaolin for a glimpse of how clay comes about in nature, in the earth. Clay minerals are not primary constituents of rocks but are what geologists call 'secondary', forming at or near the ground surface where ground and rain waters act on the existing materials. A good example is the rock granite, the major mineral of which is feldspar, rich in potassium and aluminium. Water in the ground reacts chemically to replace the feldspar with a potassium-rich clay mineral (called sericite), which then reacts further to lose its potassium, leaving kaolin.

In fact, kaolin soils are found most commonly on granite bedrock, and typically still contain a lot of potassium. Vineyard examples are the soils at Temecula Valley in Southern California, Spain's Rías Baixas and Portugal's Dão. At Larnage, in the Crozes-Hermitage region of France, the thick kaolin soils derived from the granite of the Northern Rhône not only lead to acclaimed Syrah wines but are quarried for the famous Le Panyol bread and pizza ovens.

While the relatively large clay flakes of kaolin usually give soils of adequate drainage, their chemical simplicity can lead to nutritional issues. The absence of calcium and carbonate gives acid soils that can restrict nutrient availability and curb root development. This is the case in, for example, South Africa's Western Cape, where kaolin soils are common, derived from granites such as those at Darling, Paarl and Stellenbosch.

Down on the Cape Peninsula, just south of the historic Constantia and Steenberg wine estates, there are old kaolin quarries at Noordhoek and Fish Hoek. The vineyard soils there reach acidities with a pH lower than 4 and growers routinely have to incorporate lime in the soils to achieve a balanced nutrient availability.

The other issue is the typically high potassium content arising from the clays associated with kaolin. Where the vine's uptake-selectivity mechanisms are unable to curb the abundant potassium, the excess tends to react preferentially with the wine's tartaric acid. The resulting potassium tartrate is poorly soluble and precipitates, forming the well-known white crystals, stained purple in the case of red wines, seen at the bottom of some bottles and corks. These 'wine diamonds' are themselves innocuous enough, but the resulting lowering of acidity can lead to flabby, unstable and short-lived wines.

There has been considerable excitement in recent years about a new use of kaolin in vineyards. High levels of ultraviolet radiation can disturb vine growth and sunburn grapes: parts of Australia such as the Granite Belt of Queensland can lose 15 percent of a crop due to sunburn. Sunscreen treatments in the past have had mixed results, but the new approach uses nanotechnology, and involves spraying comminuted kaolin particles each of which is smaller than a billionth of a metre. Tests have suggested that this not only blocks the ultraviolet radiation, but that the whitened leaves increase light reflectance into the canopy and improve evenness of grape ripening.

Moreover, it seems that nano-sprayed vines repel insect pests, including the sharpshooters that spread the dreaded Pierce's disease.[1] At the same time, it seems the vines optimize the way they use available water. Seeing as currently around two-thirds of the world's grapevines are cultivated in semi-arid conditions, this use of nano-kaolin could become increasingly valuable in these times of rising temperatures and concerns about water sustainability.

The miracle of montmorillonite

Moving on to the family of clay minerals known as montmorillonite – and also as smectite – we jump in complexity and difficulty. But we certainly must take a look, because it's a clay that is vital in many vineyards and can be important in the winery itself, in ways nicely illustrating the significance of those two properties fundamental to all kinds of clay.

Most montmorillonite originates from the weathering of igneous rocks like basalt and related volcanic deposits. For example, the young volcanic rocks of the Tokaj-Hegyalja region of Hungary are currently weathering into montmorillonite which, in addition to underlying some of the best vineyard sites, is quarried for industrial use. Geological processes can rework such volcanic material and re-deposit it elsewhere as sediments, so areas of sedimentary rock such as shale can also yield soils rich in montmorillonite; it's a reasonably common clay in vineyards around the world.

Now for those two properties, and first the *size* of montmorillonite flakes. They are much, much smaller than kaolin: a single flake is little more than a

1 Pierce's disease is a bacterial infection fatal to grapevines spread by sap-sucking insects such as the glassy-winged shooter. It is mainly a problem in the Americas but is spreading worryingly with the effects of global warming.

millionth of a millimetre in thickness. (That, for instance, is 50 times smaller than a single Ebola virus). The length can be anywhere between 30 and 300 times more than the thickness, so the particles are distinctly flaky in shape.

However, a major consequence of all this is not so much the unimaginable minuteness, but the surface area each flake presents. The figures are staggering. A handful of montmorillonite clay, with all its constituent minuscule flakes, will contain within it a surface area more than a thousand times greater than a handful of sand. Just a single gram of montmorillonite includes within it a clay surface area greater than a couple of parking spaces.

And here's the thing. No mineral is perfectly formed; there are always some unbonded atoms left spare at the surface, and these can be calcium, magnesium, iron, etc – the very elements that vines need as nutrients. With most minerals the number of these spare atoms is trivial but with montmorillonite its colossal surface area means that there are a great many of them, and being loosely held they are available to be detached and utilized by the vine roots. This is fundamentally where the nutrients required by the vine come from. Mycorrhizae may be involved, and in practice the nutrients are recycled through humus (Chapter 15) but the detachable atoms at the surface of montmorillonite and similar clays are the underlying source.

The second critical aspect is the behaviour of montmorillonite with water, and here it's that sheet structure again. Water can seep between the sheets, prising them apart, to be held there without actually bonding with the clay structure. Substantial amounts of water can be stored, and also loosely adsorbed onto the surfaces of the clay flakes. The benefits are obvious for vineyards on free-draining soils or prone to drought. Here are some illustrations.

The Pomerol region is famous for its clay but even so there are considerable spreads of sand and gravel where soil drainage can be excessive, especially on more elevated ground. And the highest point is occupied by Château Petrus. However, at that celebrated site within reach of the vine roots is a montmorillonite clay capable of storing water from the winter rains. In fact, the same clay comes and goes across northeast Pomerol, over a comma-shaped area that includes the vineyards of top-notch châteaux such as Trotanoy, Le Pin, Vieux Château Certan and L'Evangile, together with Saint-Emilion's Cheval Blanc and Figeac. All of them attach great importance to the presence of this montmorillonite (known locally as *argile bleue*) in their soils, at least for some of their vines.

Unlikely though it often seems to Westerners, fine wine is made in a number of Middle Eastern countries and Jordan is one case in point. Originally, Omar

Zumot used his French training to make good wines southwest of the country's capital, Amman, but he was always searching for an optimal site. Eventually, he found it up near the border with Syria, in a region known as the Mafraq. It's one of the most fertile parts of Jordan because of the basalt that once flowed here from the now-extinct Jabal Druze volcano to the north. But at an elevated place called Sama Al-Sarhan the basalt is weathering to montmorillonite clay, reducing the soil fertility nicely for vines and storing soil moisture from the winter rains. Thanks to this clay, with minimal irrigation even in this arid area, Zumot's Saint George wines now rank among the finest in the Middle East.

In recent years a number of commentators have tipped Romania as a country to watch for promising wines, and the Dealul Mare region, just north of Bucharest, as probably the best. So, around the villages of Finţeşti, Urlaţi and Fântânele, for example, several vineyards are attracting attention and they all have a certain amount of montmorillonite clay. Mihail Rotenberg was attracted to the village of Ceptura in this area to produce his eponymous wines, and especially Merlot, when he realized that the soil was similar to that in parts of Saint-Emilion, that is, underlain by montmorillonite. The area receives limited rainfall and there are some worries that the soil is too fertile, so the patches with montmorillonite soil are the most desirable.

However, there are snags to all this. Water entering the clay and prising apart the internal sheets causes the flake to swell in size: sodium montmorillonite expands to several times its original size. The swollen clay particles readily coalesce to produce a sticky mass, blocking the channels that allow water movement in the soils and hence closing down the drainage. Obviously, the facility for storing water is of little use if the water can't get out. In addition, rainwater may not be able to penetrate down into this impermeable soil, leading to surface waterlogging and undesirable puddles in the vineyard.

On top of this, there's the converse problem that when the clayey soil eventually dries out, it shrinks and cracks. The soil becomes denser, hindering root growth, and the vines respond by trying to ripen fruit prematurely. The problem is illustrated at the Cristom Vineyards in Oregon's Eola-Amity Hills AVA. The winemaker, Steve Doerner, has always specialized in producing

OPPOSITE: Autumn vines (*top*) at Cristom Vineyards, Oregon, USA, and (*below*) its montmorillonite clay soil, which is preferred here for Pinot Gris. Because the clay absorbs water, the ground becomes very sticky when wet, making vehicle or even foot travel difficult or impossible.

Pinot Noir, but he's found that Pinot Gris copes better in the block that has montmorillonite. Even so, although he had to install irrigation in order to help manage the late-season water supply, he still usually ends up picking here first.

Because of their behaviour with water, montmorillonite and related clays are sometimes referred to as 'shrink-swell', or alternatively as 'heave and subsidence' clays, and the latter phrase illustrates a further problem. If water seeps beneath a building during wet times, any shrink-swell clays in the foundations will expand and try to 'heave' the building up, albeit probably no more than an apparently trivial few millimetres. When the clay dries out the building settles back down, only for the whole process to start up again in the next rainy season.

Eventually, after several seasons of this repeated lifting and settling, cracks may begin to appear in the walls, and the structure may begin to founder. A common response is simply to fill in the cracks but that is merely masking the symptoms. To resolve the problem, either some flexibility has to be introduced into the building structure, or the construction underpinned to new foundations.

The phenomenon may be insidious but it's a very costly business. Parts of France have suffered dramatically in recent years from flooding events, but the resulting financial losses are less than those arising every year from heave and subsidence. In the United States, the costs of dealing with swelling clays on average exceeds those arising from headline-making disasters such as tornadoes, earthquakes and hurricanes. In the UK, heave and subsidence accounts for the single biggest year-on-year insurance claim for building damage, around £500 million a year. The clays, incidentally, are most common in the east and southeast of the country, exactly where vineyard expansion is currently taking place.

Furthermore, the problem is worsening. Swelling clays are rarely mentioned in discussions of climate change, but according to a recent report they pose the greatest risk to critical infrastructure, such as highways, telecoms, water, gas and electricity distribution. And all because the clay atoms are arranged in sheets.

A fine story from the Wild West

One of the two senators elected to represent Missouri in 1820, the year it became a state, was Thomas Hart Benton. A rich and colourful political career was to follow, but his overriding endeavour throughout was to promote the continued westward expansion of the United States, to meet its 'manifest destiny'. Thus, he helped author the first Homestead Act, and encouraged the founding of the state of Oregon and the annexation of the Republic of Texas.

He also championed the rights of the American Fur Company to deal in Montana, even when the operation was accused of being more interested in trading whiskey than furs. This resolute support from a politician 2,400 kilometres (1,500 miles) away was acknowledged by the Fur Company in 1846 when it completed construction in Montana of a new outpost on the uppermost reaches of the Missouri River. They named it in his honour: Fort Benton.

Shortly after that, two pioneering geologists of the American West found close by the new fort layers of volcanic ash, now largely converted to clay. They later noted that the same ashy rocks extended way south, far into Wyoming, but they decided to name the entire formation after the site of their original discovery, and termed it the 'Fort Benton Group'.

Actually, down in Rock Creek, Wyoming – five saloons but no churches; an official population of 91 including a 'tramp' – a certain William Taylor was already digging out a 'soapy clay' from these very rocks, and transporting it all the way to a cosmetics company in far-off Pittsburgh. And by the end of the century, the state geologist of Wyoming, Wilbur C Wright, was confirming that Taylor's clay indeed had special properties, so much so that he felt it justified its own name. He called it bentonite. Today, you can see sacks of bentonite in wineries all around the world.

The monetary value of bentonite to the wine industry is now huge, but it's just a fraction of its total commercial worth. In 2022, 23 million tonnes of the stuff was mined in 47 different countries, with a value approaching $US 2 billion. Nearly half is used in iron foundries (as a high-temperature bonding for moulding sand) and in pet-waste absorbents. Among numerous other things, it's used *in* pet foodstuffs, and curiously even somehow finds its way into human food. William Taylor could hardly have imagined!

We now know that bentonites around the world vary somewhat in the geological minerals they contain, but that one of them always dominates and accounts for those special properties – montmorillonite. And once again, it's that sheet structure that's important.

One use of bentonite is as a soil conditioner, with its montmorillonite increasing the water-holding abilities of soils and giving them weight. The moistened clay has the effect of 'thickening' the soil, giving it body and making it 'heavier'. And stickier. Anyone who has tramped a vineyard with montmorillonite

soils – Vaillons in Chablis, say – in the rain, will know how eagerly the sticky clumps cling on to your boots. (Just as in the hell that was the Flanders trenches during World War I; 90 percent of the Ypres clay is montmorillonite.)

However, the main use of bentonite in the wine industry is for clarification, in a process known as 'fining'. Just as montmorillonite expands with soil water, so it swells when mixed with wine, and attracts proteins and flecks of solids, the things that can cause wine to remain hazy rather than settling clear.

It can also remove excessive tannins and improve the wine's mouthfeel. In a typical procedure, at the end of fermentation bentonite granules are stirred with water and the resulting slurry added to the wine. The thin suspension adsorbs any potentially haze-forming substances and slowly settles to form a sludge, from which the wine can be racked, now star-bright as consumers expect. Some winemakers filter the wine in addition to, or instead of, fining, but filtering doesn't remove any soluble compounds that may later go on to cloud the wine.

However, the procedure of fining has its critics, both as a matter of principle – interfering with the natural progression of the wine – and through its possibly removing desirable aromatic compounds. Also, a study at America's UC Davis showed that aluminium was stripped from the bentonite during fining, particularly by high-acid wines, leading to what is regarded by some as contamination and, arguably, to possible health issues. (Which, incidentally, raises questions about the new fashion for packaging wine in aluminum cans.)

LEFT: Iron-rich clay soil desiccated and cracked during a period of drought; here in the Altai Mountains, Mongolia.

The use of bentonite in vinification has taken on a new significance in recent years, with the explosion of enthusiasm for veganism. We increasingly see on wine labels phrases like 'suitable for vegetarians and vegans', which often prompts the question: 'Aren't all wines, being fermented fruit, suitable?' Well, no, and the main reason lies in some wines being fined by using animal-derived agents, such as casein (milk protein), chitin (from crustacean shells), albumen (from egg whites), gelatin (from boiling animal parts) and isinglass (from fish bladder membranes). Consequently, vegan suitability comes about through the wine either not having been fined, or because a non-animal product, of which bentonite is the most common, has been employed.

It's perhaps somewhat ironic that biodynamic wines, with their aura of extreme closeness to nature, aren't suitable for vegans, because of their prescribed dependence on animal products – cow horns stuffed with cow manure, for instance. And as a final twist, back in Rock River, Wyoming, which first saw the commercial extraction of the bentonite now crucial to vegan wine-drinkers, the clay is no longer extracted; the business is long gone. Instead, these days Rock River's main commercial product is – beef.

Back to the nursery

You probably know vermiculite, perhaps as the stuff sold in garden centres for rooting seedlings, or from the heat insulation in your attic, but you've probably not thought of it as a clay. But it is, occurring in the ground just like other clays, though in this case with the involvement in its chemistry of magnesium, and sometimes iron, and so tending to form by the weathering of magnesium-bearing minerals such as those found in volcanic rocks. It also contains water. However, unlike the loosely attached liquid water that participates in the kinds of shrinking and swelling we've just been discussing, in vermiculite this water is bonded wholly within the solid crystal lattice: its chemical formula is written to include H_2O.

Apparently, it was known long ago in Japan that this kind of clay could be heated to produce weird little shapes, and provide amusement for children. Then, in 1824, almost certainly the same kind of stuff was dug by Thomas H Webb from the ground near Worcester, Massachusetts. He wrote that 'if subjected to the flame of a blowpipe or that of a common lamp, it expands and shoots out into a variety of fanciful forms, resembling most generally small worms having the vermicular motion exact.' So he proposed to 'term it vermiculite (worm breeder) from *vermiculare*, to breed or produce worms'.

Vermiculite was born, though the material remained something of a curiosity until the early years of the last century. In 1923 the first commercial mine was started in Libby, Montana, to supply the construction industry. Today, the clay is mined all over the world and industrially processed. In particular, the extracted material is cleaned and then heated very quickly to force the internal water turn to steam, which blows apart the mineral sheets to give the now-familiar accordion-like form. Having been expanded in this way, the clay is clean and light to handle, odourless and mould-resistant, and has a remarkably low density that makes it great for heat and noise for insulation.

And for those same reasons it makes an ideal material for embryo plant roots, a 'rooting medium' in horticultural speak. Thus, gardeners can buy bags of it for raising plant cuttings, and if you've ever bought already established plants in little pots you might have noticed bits in the soil sparkling in the sunlight – they're vermiculite. The kinds of temperatures employed in its production – upwards of 100°C (212°F) – mean that situation has no direct relevance to vineyards, but all vine nurseries employ expanded vermiculite (or another mineral product called perlite). Generating new vines is an aspect most of us rarely think about, but these days it's a vital and highly specialized business. The latest technologies involve sophisticated *in vitro* micro-propagation of vines, but vermiculite is still required, to bring on the seedlings before planting out in the vineyard.

So, we've seen the importance of clay in the wine world as well as glimpsing some of the wide range of industrial applications. But clay might also be of relevance to individuals, seeing as – we are told by promoters – applying or ingesting clay 'fuels creative passion', gives hair 'a less-defined, sleeker finish', 'promotes normal stools', 'detoxes armpits', and so on. One advertisement for a facial treatment based on clay declares that: 'You can't buy happiness but you can buy clay, and that's pretty much the same thing.' Clay is certainly wonderful stuff.

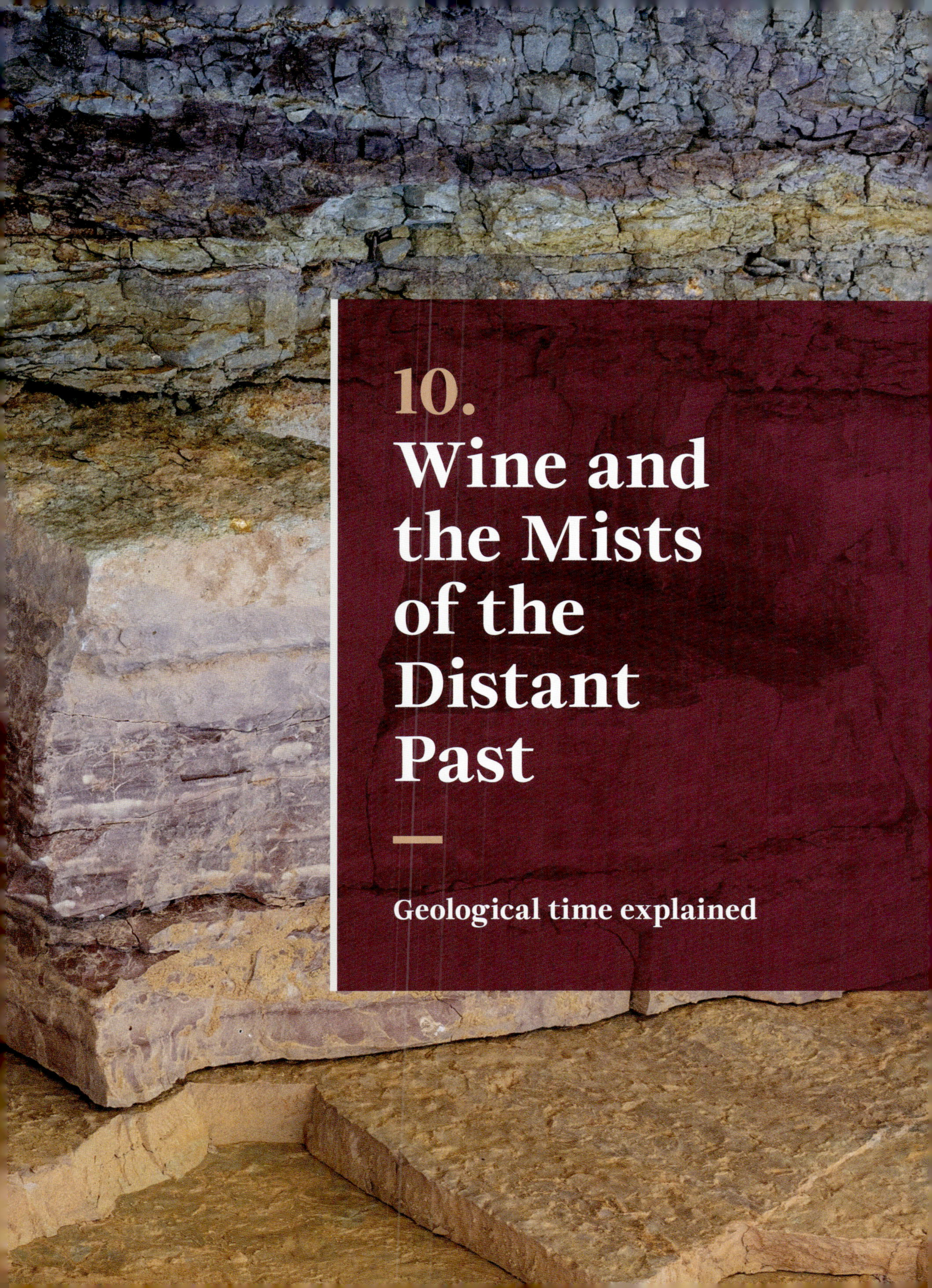

10. Wine and the Mists of the Distant Past

Geological time explained

Wine and the Mists of the Distant Past

'The area's best vineyards are on *Turonian* soils.' 'The wine's secret is the grey-blue *Devonian* slate.' 'Our Shiraz grows in soils *500 million years* old.' '*Cretaceous* limestone is best for our vines.' In these quotations, what do all the words that I have italicized have in common? Answer: they are all to do with geological time.

It seems that many in the wine world find these kinds of words important and enjoy using them. Perhaps part of the appeal is that they carry an air of mystery, as they are evoking far distant times. However, I do sense that many writers are not entirely comfortable with them. They are, after all, technical scientific terms that are dealing with tricky concepts. And in any case, how on earth do geologists know these things? How do they know that one rock is older than another one that looks pretty much the same? What do these geological time names actually mean, and how are those inconceivably long ages determined? In this chapter, I attempt to shed some light, in order to provide a context for the wine enthusiast.

But first, I should state my own position. I cannot see – to put it very bluntly – what, apart from passing academic interest, all this has to do in practice with vines and wines. The words are referring to the geological age of the vineyard bedrock and surely the vine doesn't care, so to speak, how long ago that happened to form. By definition, the bedrock is pretty much intact, apart perhaps from some fissures that may conserve supplementary water into which deep roots tap: it is the overlying loose material, the soil, that is largely used for water and nutrition. And almost invariably the age of the soil will be *unrelated and vastly younger than the bedrock*.

The age of the soil is certainly relevant, as it is continually changing on a human time scale. To stretch a point, every time a grower walks his vineyard, every time a machine drives over it, some of the soil's properties will be at least fractionally affected; every rainstorm, every harvest will modify the chemistry.

But it is not the age of the soil we are talking about here. Quotations like those above, even if they say 'soils', are referring to bedrock, and how many hundreds of millions of years ago it formed would seem to be of no practical relevance.

Nevertheless, be that as it may, the fact is that these geological words and concepts are used very commonly, in all kinds of wine literature, and so I would like to assist in their being used correctly and with understanding.

How geological time works is perhaps best understood by recalling how it was 'discovered'. So, in outline, this is the story.

The dawn of geology

Modern geology began two or three centuries ago, essentially when it dawned that questions about the physical world were better answered by going out and observing nature rather than poring over ancient scriptures. Very soon, once this particular penny had dropped, various 'principles' were being enunciated. They were given noble titles, though they seem pretty much self-evident to us now.

For example, the 'principle of superposition' tells us that (just like with pouring sand into a bucket) in a pile of sediment layers, the lowest was deposited first – it is the oldest – and the layers get younger upwards. The 'principle of original horizontality' observes that all sediments are originally laid down in horizontal layers (even on a hillslope, because Earth's gravity works vertically downwards), so that if we see inclined layers the tilting must have happened later. The 'principle of cross-cutting relations' says that a particular geological feature which cuts across another one must be the younger of the two; the other one must have been there already in order to become cross-cut. With such armoury, the early geologists were in a position to look at, say, some rocky cliff and work out a sequence in which things happened, even if it was in the distant past.

Less straightforward was linking this sequence with that seen at another place, but even the pioneers had ways. For example, if a particularly distinctive feature was involved, say an unusual volcanic ash deposit, it could act as a marker. The white, powdery, fossil-rich chalk seen in southern Britain – the foundation of numerous English vineyards – is a very characteristic deposit. The conditions on the sea floor thereabouts at the time of chalk formation were very special. The hardened and uplifted rock now makes the famous White Cliffs of Dover and the English coast around them, but there are very few places in more distant parts of the world where it occurs. It is distinctive.

However, chalk does appear on the other side of the English Channel in the cliffs around Boulogne, so it seems reasonable to infer that they were physically joined until the Channel was eroded down through them. In France, this same distinctive chalk formation stretches south-eastwards, all the way into the Champagne region and beyond.

So although the occurrences today are spread and not continuous, geologists correlate them. Also, the particular fossils the chalk contains are different from those in the strata below, which are different from those above. Elsewhere this same order of fossils is found even where the middle formation is not chalk but some other sedimentary rock. The inference is that in such places the conditions on the sea floor were not suitable for chalk deposition, but the material is of the same geological age as the chalk.

Using these kinds of principles, the early workers were soon able to define past intervals of geological time, and they gave them names. The kind of scheme that resulted is the foundation of the geological time scale we use today (*see* chart on page 169). Many of the names came from the places where good evidence was first recognized: Devonian from Devon, Permian from Perm, Jurassic from the Jura, etc. Cambrian came from Roman attempts to render in Latin – *Cambria* – the local Celtic name for the land today called Wales: *Cymru*. The pioneers realized that these Cambrian rocks were very old, but also that there were even older ones, which were exceedingly difficult to disentangle. So they were simple shrugged off as 'Pre-Cambrian' (note, for the moment, the hyphen).

This approach of inferring that one thing is older than another is called *relative dating*. There are no numerical dates or times involved. It's like historians saying that the Renaissance came before the Baroque but without giving any years. This new demystifying of the Earth's history thrilled the public in Victorian times and elevated geology to a heroic science. In line with this, at the 1851 Great Exhibition in London the grounds of the Crystal Palace incorporated a huge representation of the newly conceived geological time scale, complete with oversized models of fossils. In the hollow body of the *Iguanadon* a celebratory fine dinner was staged. ('Sherry, port, moselle and claret' were served). The geologists of the time could be mightily proud of their achievement.

But there was a big problem. Something was glaringly missing. *How old* were these geological periods? *When* did they happen and *how long* did they last? No one knew, and there was no way of finding out. I imagine that those pioneering geologists would have given their eye-teeth for a way to put some numbers on their time scale.

The discovery of rock dating

Geologists call the business of giving a numerical age to a rock *dating* it – source of many an undergraduate quip about geologists loving their subject. The approach is properly called *numerical dating*. It's like art historians, continuing the analogy mentioned above, saying that the Renaissance started about 700 years ago and the Baroque period about 400 years ago. Ingenious ideas were proposed on how this could be achieved in geology, such as counting the growth rings on fossil corals, or estimating the rate of supposed increasing salinity of the oceans or of Earth's inferred cooling from a primeval fiery state. Unfortunately, there was a suspicion even at the time that the bases of these procedures were mistaken, and we now know that to be the case.

ABOVE: Fossilized shells (ammonites) in a very fine-grained limestone, from Croix-de-Saint-Concor, Savoie, France. Geologists can deduce, using relative dating, that the limestone was deposited in early Tithonian (Jurassic) times. Numerical dating shows that this was around 147 million years ago.

Another seeming flaw with these methods was that they gave an age for the Earth in many *millions* of years. That struck even some geologists as unthinkably long. In the centuries preceding all this, various clerics had attempted to work out the age of the Earth from translated scriptures, and though estimates varied none of them exceeded 10,000 years. In the English-speaking world a particularly influential 'computation' (*sic*) based on interpreting a number of Christian and other texts pinpointed the origin of the Earth as 10.30 in the morning of a particular November day in 4004 BCE. It became something of a dogma. But with the emergence of geology, it was soon realized that this gave nowhere near enough time to explain the observed features of the world. Few, however, conceived of many millions of years, let alone the vast periods known today.

Eventually, bit by bit, a method emerged for dating rocks that was sound and reliable. But it was very slow going. Radioactivity was discovered in the late 1800s, and it took the first decades of the 20th century to gradually elucidate its nature. Strings of Nobel Prizes ensued as it was demonstrated that certain natural elements were intrinsically unstable and through time gradually *decayed*, that is, threw off some of their constituent parts – radioactivity – to attain a more stable arrangement.

Understanding of the physics became ever more sophisticated and it became clear that the rate of decay of any given radioactive element was fixed. Not just roughly constant but an unchanging fundamental property, no more variable than, say, the speed of light. Before long, there was analytical equipment that actually allowed the behaviour of short-lived radioactive elements to be tested and measured, and it all agreed. No matter how extreme the conditions, the rate just could not be made to change.

And so it dawned that if we could analyse the proportions of decaying matter in rocks then, knowing the rates, we could work out how long the process had been happening, that is, the age of the rock. In, for example, a rock solidifying from a hot, lava-like melt, the radioactive materials get locked into the crystal lattices of minerals, so today we can measure their proportions and derive when the process began. (These decaying elements and their products are different from those not involved with radioactivity and so they are distinguishable. Technically speaking, they're different isotopes.)

Theory and laboratory observations agreed, the principle was simple and elegant; the main drawback at the time was the clumsiness of the analytical equipment. And the legacy problem persisted: when this method was tried, the rock ages came out in *hundreds of millions* of years. It was such a mental wrench from what had been believed before.

But as the technology improved, as more laboratories around the world came up with the same ages, as the same rocks repeatedly gave the same result and relatively younger or older ones gave corresponding absolute ages, the strictures of earlier thinking just had to be discarded and the method accepted. And for geologists, this nicely gave plenty of time to incorporate all the natural processes they dealt with, even the excruciatingly slow ones. It all fitted together. So, by the mid-20th century a convincing geological time scale complete with numbers had been established.

Much of the work since then has been refinement and tackling those rocks that are hard to date, either because they have very little radioactive material or, for various reasons to do with their history, give values that are hard to interpret. Equipment has improved unimaginably such that all sorts of radioactive elements can now be analysed and, if any further corroboration of the method were needed, there are now a dozen or so ways of dating rocks, based not on radioactivity but on quite different physical principles. Each of these is restricted to certain kinds of rocks or certain periods of geological time, and so lacks the power and versatility of the radioactive method, but they all give results that agree.

Immense but manageable

The enormous numerical ages we now recognize are beyond our comprehension. But that, as I plead with my students, is no reason for us to be cavalier with the numbers. If we are to understand geological time, we have at least to get the orders of magnitude right, the exact numbers of zeroes. And I'm afraid that numerous writings, even responsible ones, have it all wrong.

I wonder to students if they would be so carefree with the zeroes if it was money, say your first prize in the Euro-lottery or the hundred-billion dollar Zimbabwean banknote in your hand as you attempted to buy something. In Croatia today you can buy a bottle of wine for five euros or seven US dollars; in the 1990s you'd have had to count up the zeroes on a banknote such as that in the picture overleaf (and then realize it wasn't enough).

As technology improved it became possible to date even the oldest rocks on Earth and geologists started talking about *billions* (thousand millions) of years. It turns out that the Earth formed about 4.6 billion years ago. One thing this means

LEFT: Large-denomination banknotes in Yugoslav dinars (*top*) and Zimbabwean dollars (*bottom*). Daunting though the very large number of zeroes is, they are critical for purchasing power, in the same way that the exact number of zeroes in geological time is essential to proper understanding.

is that the greater part of the planet's history is occupied by what is now called the *Precambrian* (note now the lack of hyphen!). It occupies more or less the four billion years of Earth's age, which means the remaining fractional part – the 0.6 billion, or 600 million years – covers pretty much all the geological time we talk about in vineyards (*see* time scale chart on page 169). There are vineyards on bedrock a billion years or more in age, such as at Stellenbosch, in South Africa, in Western Australia, and a number of US states, including Virginia and the Texas Hill Country, but it so happens that most are on geologically young bedrock.

Not uncommonly, descriptions of the geology of a vineyard say that it formed in past 'millennia', which really means just thousands of years. Or it formed back in 'prehistory', which properly means the time of humans before written records began, ie, just a couple of hundred thousand years or so. However, in general, the geological age of vineyard bedrock is measured in *hundreds of millions* of years, or more. If our understanding of the geological ages of vineyard rocks and soils is to make any sense, we have to have the order of magnitude right.

The geological time scale

So, the coming together of the relative and absolute dating approaches has given us the geological time scale (as illustrated on the next page). Its framework is by now very well established, and it provides an important working calendar of the Earth's history. But there remain plenty of terminological problems in detail, and one difficulty that impinges on the wine world stems from the plethora of names that the geological pioneers established but which are now no longer used in geology.

The early names were often parochial, with no obvious correspondence to names elsewhere. Moreover (as though not to make things too easy!), the early geologists liked to modulate the name a little, such as with *Toarcian*, which comes from Thouars, in the Loire Valley, and *Coniacian*, derived from the city of Cognac. They also liked to show off their knowledge of classical times. Thus, the *Zanclean* stage is named after Zanclus, the pre-Roman name for Messina on Sicily; *Callovian* is from the Latinized name for Kellaways Bridge, near Chippenham, in Wiltshire, England. I have seen all of these names in wine writings. How many modern wine drinkers know the Latin name for Bordeaux? It's Burdigala and gives its name to the *Burdigalian* stage; the *Lutetian* comes from Lutetia, the Latin name for Paris.

But it is the parochiality of many of the names that is problematic in these days of global communications. And so, for some decades now, a thoroughly international geological commission (http://www.stratigraphy.org/) has existed, which has a formal and quasi-democratic system of agreeing which of the local names should be discarded and which adopted internationally, in the interests of regularizing and harmonizing usage across the world. The commission also attempts to identify and agree where in the world the rocks definitively representing each named interval exist. As it turns out, rarely do these most useful localities coincide with the places that spawned the name. Thus, for example, the *Ypresian* stage, which long ago took its name from Ypres (Ieper) in Belgium, is now defined in rocks near Luxor, Egypt; the name *Turonian* (which also finds itself as the name of a Vouvray wine) came from Tours, in the Loire Valley, France, but the agreed reference rocks today are near the town of Pueblo, in Colorado, USA.

By and large, the commission's recommendations have been smoothly absorbed into geology but, wholly understandably, they have proved slow to trickle into the wine world. The upshot is that a number of terms commonly seen in wine writings have a meaning that now diverges from its correct usage in geology, or, indeed, have become technically obsolete. A few examples of these instances follow. All the terms are seen in wine literature, and all clearly are serving a purpose there, but in different ways they are at variance with the modern usage in geology.

Helvetian, Tortonian and Serravalian in Barolo

In the Barolo region of the Italian Piemonte, growers have long recognized that the wines produced from the west of the area tend to diverge in character from those produced in the east. The difference is often ascribed to the bedrock of the two areas, which varies somewhat in nature, and it is usually its differing geological *ages* that are highlighted. Thus, we read that the rocks of the western area are Tortonian whereas those in the east are Helvetian. Without extra knowledge, these names simply indicate the time at which the rocks formed, nothing more, but my purpose here is to illustrate the localized name problem, and how things evolve with improved geological knowledge.

Barolo has an intricate history of different local geological names being employed, and way back in 1865 the term *Serravalian* was proposed for the time at which the rocks in the east of the area formed. The name came from a municipality right there in the Piemonte region – Serravalle Scrivia. But it never caught on. *Helvetian* became preferred instead, even though that name obviously derived from relatively distant Switzerland. So Helvetian became the established term, and geologists used it in Piemonte and elsewhere right up until 1957.

Then they discovered that it was an inappropriate name, because the Helvetian rocks in Switzerland turned out to be equivalent in age to the already established Burdigalian stage! 'Helvetian' had to be discarded. But the rocks in eastern Barolo weren't Burdigalian, so what to call them? The old name Serravalian was revived and it is now the official geological term for this particular geological age.

Actually, the reference rocks for the Serravalian are now defined in southern Malta, but the point here is that although the word 'Helvetian' is much seen in Barolo wine literature today, for over half a century the term has been obsolete in geology. So technically speaking, to be in line with geological usage, wine authors ought to be using the term Serravalian and not Helvetian.

Geological Time Scale

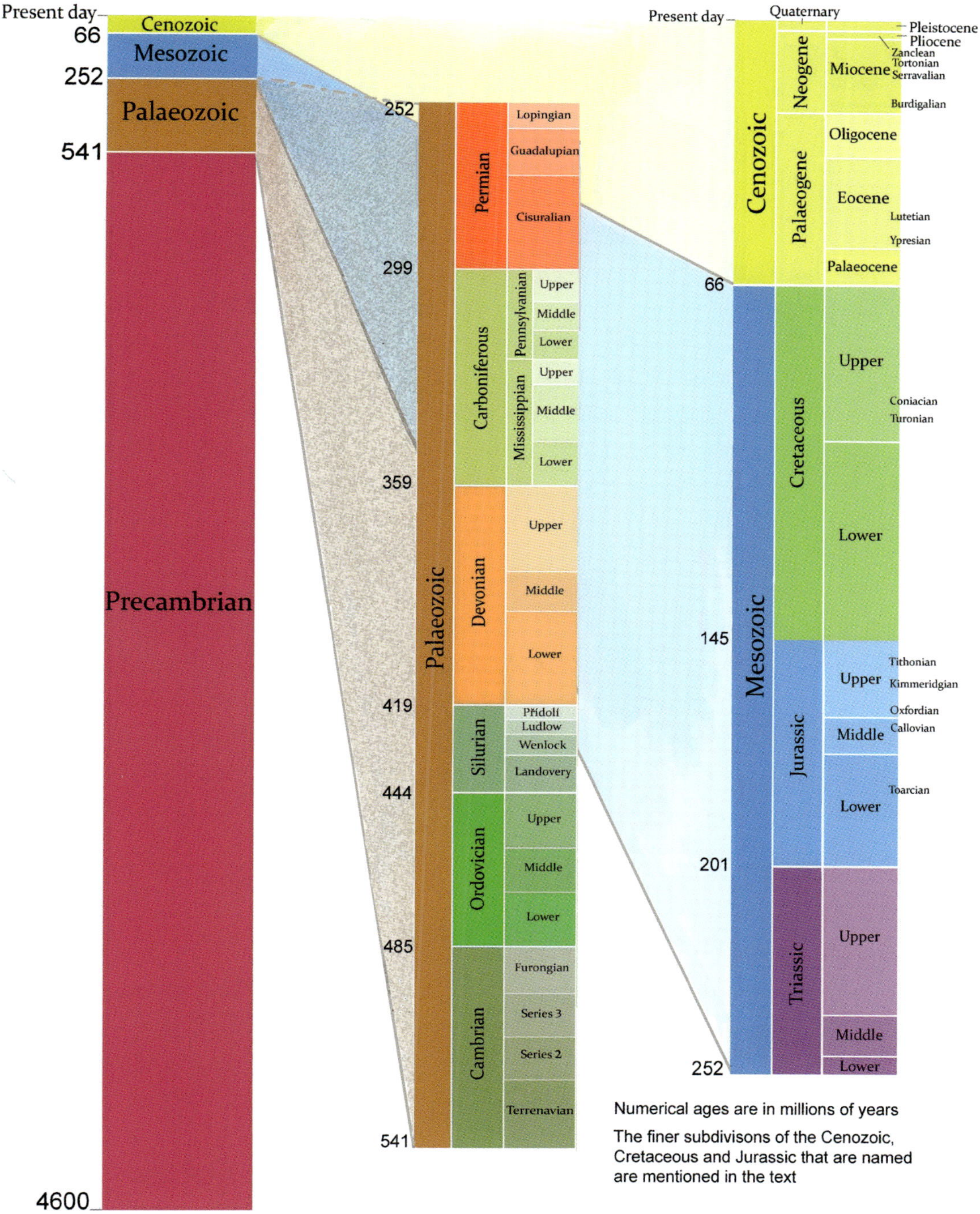

Numerical ages are in millions of years

The finer subdivisons of the Cenozoic, Cretaceous and Jurassic that are named are mentioned in the text

Muschelkalk

This term is often used in the wine world for a period of geological time – and once it was. These days in geology, however, it is a subtle blending of time, place, fossils and rock type.

When relative time was being explored in the early days of geology, it was in Germany that a Triassic period was defined, named because there it contained three clear divisions. The middle one was called Muschelkalk. The other two names, Bunter and Keuper, have disappeared from the international time scale though they do occasionally appear in vineyard writings. Muschelkalk is also officially obsolete in geology, but is still used fondly in Germany and neighbouring countries by wine writers, and appears on several wine labels.

The various rock types that comprise Muschelkalk are very largely calcareous, and replete with distinctive fossilized seashells, hence its German name for 'shell limestone'. The setting in and around Germany in which these creatures lived and the sediments were deposited was very localized and so it proved difficult to correlate them with rocks elsewhere. Consequently, although the name Triassic has been retained internationally, for many years the definitions of its subdivisions have been based on rocks from elsewhere in the world.

Thus, the term Muschelkalk is only used today in northern and central Europe, for the unusual calcareous rocks that formed there during Middle Triassic times. And for the vineyards in that region. For example, Muschelkalk underlies parts of southern Pfalz, Rheinhessen and central Franken in Germany, and Neusiedlersee-Hügelland in Austria's Burgenland. A number of Grand Cru vineyards in Alsace broadcast the fact that they are sited on Muschelkalk. But it is geologically wrong to say that the rocks formed in Muschelkalk times, and away from this region the term is little known.

Primary rock

We have just seen a number of time-terms that are still used in wine writings despite being outdated in geology, but surely the record for this must go to the term 'Primary rock'. Vineyard descriptions from Austria, and especially those from the Wachau and Kremstal areas of the Danube, enthusiastically mention the word *Urgestein*, usually translated as *Primary rock*, or occasionally as *Primitive rock*, and that it dates from the Earth's '*Primary era*'. However, in geology these terms – in both German and in English – and the concepts they imply have been obsolete for around a couple of centuries! Austrian geologists have long been trying to discourage use of the word; geologists elsewhere would just be baffled by it.

The geological materials in those Danube vineyards consist principally of relatively recent loess and alluvium, which contrast with a range of older, very much harder rocks that includes amphibolite, slate, basalt, granite, gneiss and schist. These comprise the Primary rock. So with prior knowledge the term can provide a convenient shorthand summary label for that wide range of rock types.

The problem is with the word 'primary'. It carries an implication of great antiquity, of something special, elemental, primeval, and some go on to believe that this brings something special to the wines. (Actually, with ages less than 500 million years old, little more than a tenth of the Earth's history, these rocks are, geologically speaking, not particularly old.)

ABOVE: A wall constructed of blocks of dark amphibolite and gneiss – so-called Primary rock – next to the Achleiten vineyard, Wachau, Austria.

In the dawning days of modern geology there was a theory that the primordial Earth was entirely covered by a lifeless ocean, from which the metamorphic and igneous rocks were *chemically precipitated.* These were the Primary rocks, defining the Primary era of the planet's history. Then, after life had appeared in the ocean, sedimentary rocks formed and contained fossils: these were Secondary rocks, from the Secondary era. (Tertiary and Quaternary divisions were added later.)

This thinking was widely admired and influential at the time, not only in its Saxon heartland but by some in Britain and other parts of Europe too – but only briefly. For in addition to the physical snags of such implausible volumes of ocean water, incompatible geological observations rapidly accumulated. Igneous rocks were demonstrated to have once been molten and not precipitated, metamorphic rocks to be changed variants of rocks that went before, and, crucially, instances of both were reported that were demonstrably much younger than fossiliferous rocks. (That's why today, 'igneous' and 'metamorphic' have no age connotations.) Thus the notions of a primeval ocean, a Primary era and Primary rock rapidly disappeared from geology.

Age of bedrock versus the age of soils

There seems to be an unwritten understanding that the adjectives in terms such as 'granite soils' or 'limestone soils' don't literally describe the loose material of a vineyard but refer to the intact parent rock from which the soil was derived. And they do convey some general information on the likely nature of the soils.

But where time adjectives are involved, things can get confused. For example, descriptions of vineyards located on Precambrian bedrock enthuse about how primeval the soils are; others tell of the importance for the vines in the Middle Mosel, Germany, of the Devonian soils, and in the Côtes de Provence, France, of the Permian soils. These soils, however, as opposed to the bedrock, are not of these ages: it's all a bit misleading.

Much-travelled materials such as loess and alluvium[1] – and these account for many of the world's vine-growing regions – have no connection with the bedrock on which they happen to rest. And even in areas where the soil particles have

1 Loess is fine-grained, wind-blown sediment (covering up to 10 percent of the world's land area). Alluvium (loosely packed clay, silt, sand and gravel) is sediment deposited by running water.

barely moved, the age of the soil itself is quite independent of what geological age the parent bedrock happens to be.

For example, promotional literature on the Heathcote wine region in Victoria, Australia, is fond of enthusing about the vast antiquity of its 'Cambrian-age soils', some even asserting them to be the oldest in the world. (In fact, most of the Heathcote bedrock is geologically younger than Cambrian, apart from two narrow strips, and even these are considerably younger than, say, the 1,500-million-year-old bedrock of parts of the Margaret River area.) But the point here is that the Heathcote *soils* are not of Cambrian age and are probably a few millions of years old at the most. And most soils are geologically very much younger than this. The Gimblett Gravels of Hawke's Bay, New Zealand (*see* Chapter 12), were forming until little more than a century ago!

The rate at which soils form from bedrock depends greatly on climate and a host of other things, so the numbers vary greatly. Some workers estimate that it takes around 100 years to generate two to five centimetres (one to two inches) of soil from volcanic ash in the warm, humid tropics, and 5,000 years to produce one centimetre (half an inch) of soil from hard rocks in cool temperate climates. One commonly quoted global average value is 178 years to develop one centimetre (half an inch) of soil. Of course, whichever number we take, to us these are slow processes. But even so, they mean that soil ages are typically measured in thousands and perhaps a few millions of years, as opposed to the hundred million-year ages of much vineyard bedrock.

Moreover, as the soil develops it undergoes erosion in one way and another. For example, whatever soils existed before the 'Ice Age' in places such as northern Europe, Tasmania and New Zealand were stripped away by the ice sheets of those times, so the soils there today are no more than 12,000 years or so in age, at the very most.

The actual age of a soil is difficult to establish, not least because of its ongoing evolution (and the radioactive method won't work here), but we can talk very generally about young, mature and old soils, or perhaps about the stage of compaction or nutrient leaching. Such aspects of soils are very relevant to viticulture. In other words, the age and properties of the soil are far more meaningful in vineyard descriptions than proclaiming the distant and unrelated geological age of the underlying bedrock, even if we understand how geological time works.

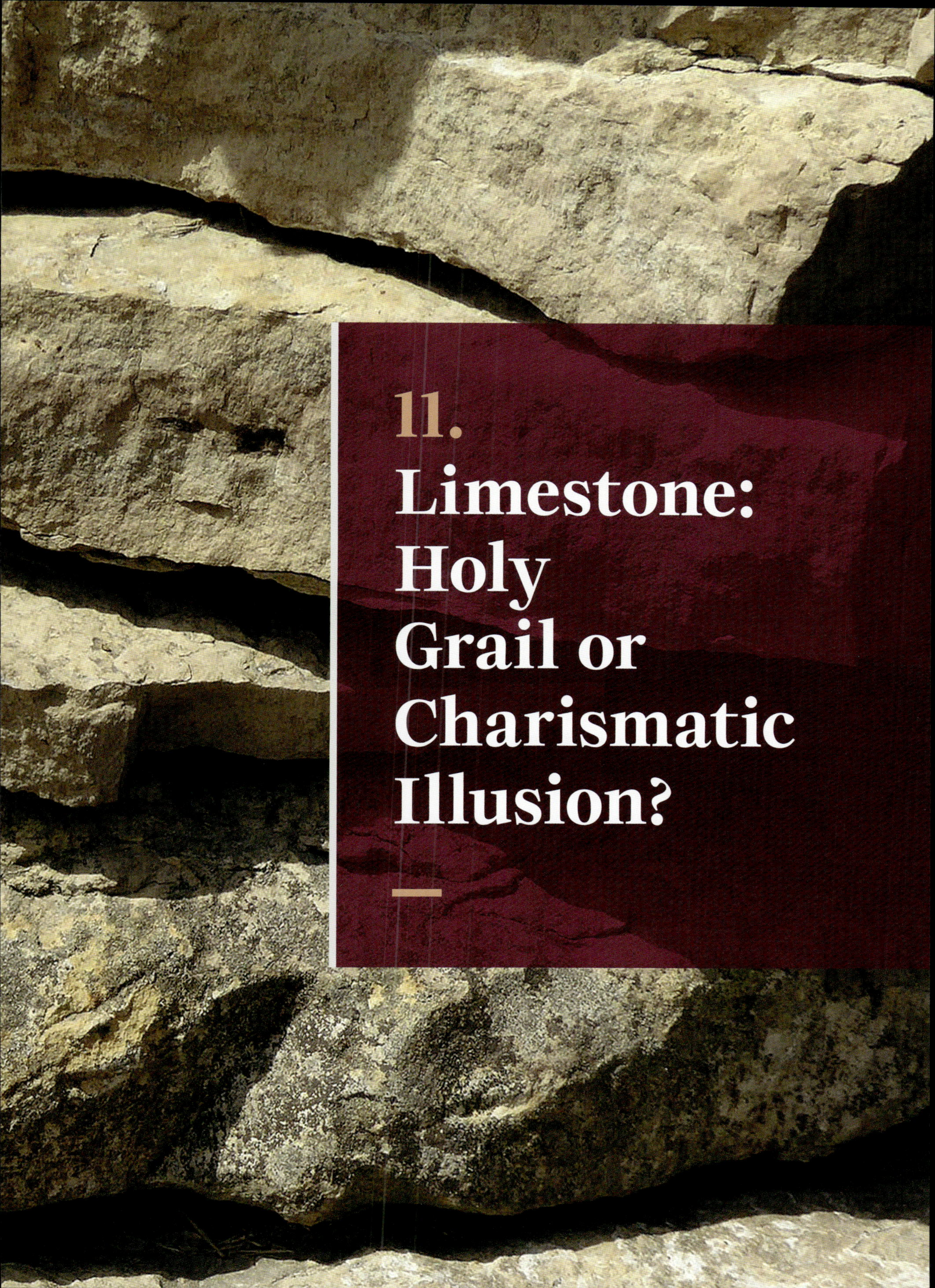

11. Limestone: Holy Grail or Charismatic Illusion?

Limestone: Holy Grail or Charismatic Illusion?

In classical times the goddess Tethys and her consort Oceanus were believed to watch over all the waters of the Earth. (And Tethys also found time to raise more than three thousand junior goddesses.) In modern times she lives on in the name geologists give to a mighty ocean that once stretched over halfway round the Earth. The western reaches of this vast Tethys ocean were shallow, warm and sunlit – as if today the Bahamas extended from what is now the British Isles to the Himalayas. The waters offered just the conditions for a flourishing marine life, including all manner of shellfish, copious plankton and widespread reefs of clams and corals. Once they'd lived their lives their shells, made of calcium carbonate, settled in thick deposits of calcareous detritus that accumulated on the sea floor.

What does all this have to do with wine? Well, with time the calcareous deposits hardened into rock – limestone – and Earth's internal forces thrust the sea floor upwards to make land, the very land where eventually wine-yielding grapevines were to originate and flourish. In the Eastern Mediterranean–Caspian Sea region, where *Vitis vinifera* began, there's a huge amount of limestone, and as the vines spread westwards they encountered yet more of it, together with its calcareous variations such as chalk and marl. About half of France is underlain by limestone, and nearly a third of Italy and Spain. In other words, thanks to the Tethys Ocean, limestone soils are prominent in the original milieu of the Eurasian grapevine.

And so when in more recent times certain areas were identified as producing superior wines it was almost inevitable that these kinds of soils would be involved.

PREVIOUS PAGES: Slabs of a limestone rich in magnesium called dolomite that give their name to Flat Rock Cellars, Jordan, Ontario, Canada.

And become famous. In the wine world, the marls of Chablis, the chalk of Champagne, the limestone of Meursault, etc, are legendary, so much so that limestone and its relatives are seen by some as the dream soil for viticulture. Burgundians rejoice in their good fortune of having calcareous soils; in the New World, where limestone is relatively scarce, some growers passionately seek it out: 'we pounced on this area because of its limestone, for wine the holy grail'; 'like the holy grail, limestone has long been sought out by viticulturists and winemakers'.

Wine writers and promoters love limestone. The listings of one New York wine merchant mention the word over a hundred times. Some wine descriptions use it to signify quality: 'the more limestone is present, the finer the wines become'; 'limestone is what gives this wine its soul'. Some commentators actually taste the rock in wine: 'the limestone rolled onto my tongue'; 'you can smell and taste the limestone'. All this is helped by our easy familiarity with the rock.

The friendly rock

As it happens, these same Tethys-born lands would also spawn Western civilization, which soon made full use of its ambient limestone. The pyramids of Giza and the great temples of Gozo are made of limestone, as is the Parthenon. From more recent times, notable American buildings such as the Lincoln Memorial, the Empire State Building and the Pentagon, together with well over half of the US state capitols, are built of limestone. So are many celebrated European monuments, including many of the great castles and cathedrals. In London, Buckingham Palace, the Houses of Parliament and the National Gallery are made of limestone.

How many poems (as in WH Auden's *In Praise of Limestone*) and novels have been written about a particular kind of rock? Fiona Farrell's book *Limestone* (2009) opens with 'I was born in limestone country. This has implications. It is a different thing to be born to clay, or schist, or basalt….' *Limestone Country* (2017) by Fiona Sampson begins 'This book is about a love affair with limestone' and goes on to remark that 'limestone is welcoming to humans, who have long found it adapts easily to their needs, as they to it'. Limestone even has a part in Christian ritual. At the Feast of Epiphany, chalky limestone is blessed by a priest: 'O Lord God, bless this creature chalk to render it helpful to men.'

So, we are comfortable with limestone; among cryptic rocks like arkose and aplite, gabbro and greywacke, it seems like an old friend. But does that make it a holy grail for vine growers? Of course, superlative wines come from many kinds of soils but all of the above illustrate that limestone is held in particular regard. It seems to have charisma. And certainly if one rock were to be singled out as possibly being special for viticulture then it would have to be limestone – because it is *different*.

A different kind of rock

Most rocks and soils are based on the two most abundant chemical elements in the outer part of the Earth, silicon and oxygen. They are *silicates*, with the two elements being held together by a handful of common metal elements. In fact, although such rocks as schist, granite, gneiss and shale have different geological origins they are chemically remarkably similar. They all weather down to give soils with similar ranges of water-holding and drainage properties; they all potentially yield the same full range of mineral nutrients for the vine. Just as with, say, volcanic rocks and soils, none has any special ingredient. Neither does limestone, but in its chemical and physical properties it's different.

Pure limestone is made of calcium carbonate, $CaCO_3$. Chalk is one kind of limestone, rich in the fossilized remains of a specific plankton; marl is an intimate mixture of clay and limestone. All are referred to as being *calcareous*. The key feature is the carbonate, the CO_3 part.

We all know in these days of concern about greenhouse gases and climate change that the air around us contains carbon dioxide, a gas that amongst everything else dissolves in the Earth's waters. The resulting weak acid reacts readily with calcium carbonate to form a compound that only exists in water: calcium bicarbonate, $Ca(HCO_3)_2$. In other words, limestone *dissolves* fairly easily, in contrast with the virtually insoluble silicate rocks. At the same time, a second feature of this compound is its incorporation of hydrogen, which acts to reduce acidity. (Strictly speaking, acidity is due to hydrogen being freely available for chemical reactions; the pH scale, arranged negatively, is its measure.) The result is that soils derived from limestone tend to be alkaline, that is, with a pH of 7 or more.

A rock that dissolves

First, the significance of limestone dissolving in water. Well, it led Auden to write: '*If it form the one landscape that we, the inconstant ones, Are consistently homesick for, this is chiefly Because it dissolves in water.*' Limestone can be a physically

LEFT: *Garrigue* – the aromatic shrubland that develops on limestone in Mediterranean regions. Its aromas can be transmitted through the air to grape skins and stems so may influence wine flavour. Mont Saint Victor, Fontjoncouse, Corbières, France.

tough rock, commonly resisting erosion to give upland areas, and this coupled with its chemical solubility leads to distinctive and striking landscape features. Examples are the strange, tourist-attracting shapes of the 'Enchanted City' near Cuenca in Spain, the magnificent fluted cliffs at Cassis in France, and the channelled escarpments that tower above vineyards in the Alps and the Jura.

Plateaus of limestone are well known in the wine regions of southern France, as is the heady mix of herbs that grow in their dissolved fissures, the *garrigue*. The word is a favourite in descriptions of wine aromas: 'wine with scents of the *garrigue*'; 'expressing notes of *garrigue*' and the like. (The equivalent word for aromatic shrubs growing not on limestone but on silicate rocks is *maquis*. When did you last see *maquis* in a wine tasting note?)

Most of the world's caves have formed in limestone, where underground rivers have dissolved away some of the host rock. Some natural caves have been expanded over the centuries as underground quarries for limestone building blocks, only to become disused and now adapted for wine storage. Putting aside

claims that the walls of these cellars somehow improve the bottled wine – 'you can taste these limestone walls in the glass' and so on – the great virtue of caves is their provision of ideal conditions for storing and maturing wine.

Famous are the limestone wine caves below Reims, Saint-Emilion and parts of the Loire Valley in France; in Spain's Penedès, Rioja and Ribera del Duero; and the extensive ancient cellars below Chateau Ksara in Lebanon. Less well known are those north of Moldova's capital, Chişinău, reportedly the world's largest. The tunnels at Milestii Mici stretch for no less than 240 kilometres (150 miles) and contain the world's largest collection of wine bottles (more than 1.5 million according to *The Guinness Book of World Records*). At nearby Cricova, you can drive your car along the 100 kilometres (60 miles) of limestone tunnels, bearing names such as Cabernet Street, Muscat Street and Aligoté Street.

The solubility of limestone means that any fine particles in soil are soon dissolved away, so limestone soils tend to be coarse, stony, and very well drained, ideal for moist northern European areas such as England and Champagne. At the same time, some limestones contain abundant little gaps between their constituent calcite grains, some enlarged by dissolution, and this allows storage of water. Such a balance between water drainage and storage, particularly well shown by some chalks, is ideal for viticulture. It's why the chalk strata beneath London and Paris are such important aquifers for those cities.

The sweet earth

Second, the alkalinity. Or more precisely, reduced acidity. European farmers have known for millennia that vines grow well on calcareous soils: Pliny remarked that for vines in Italy 'the chalky earth of Alba Pompeia is preferred'. The Romans of the time tasted water that had percolated through baskets of soil in order to gauge whether the soil was 'sweet' or 'sour'; they knew that the latter could be 'sweetened' by spreading lime. Pliny reported that the tribes of modern-day Burgundy added ground limestone to their vineyards and that farmers in Britain had mastered the process of spreading marl to sweeten and improve their lands.

Marl spreading remained important in England. When Arthur Young of Bradfield, Suffolk, lost his wholly unloved wife the only tribute he could think

PREVIOUS PAGES: View of the Acropolis, a hill of limestone in Athens, Greece. The monuments, such as the Parthenon and Erechtheion, are largely made of marble on a limestone base.

of to have engraved on her memorial was that her great-grandfather had used marl on his fields. Numerous texts on liming were published, giving precise recipes and instructions. However, they floundered when it came to explaining how it worked. In 1675, John Evelyn felt that the 'fecundity of the earth' was improved by lime because it contained the 'alchemical spirit' of nitre (the basis of gunpowder). By 1764, the chemist Johan Wallerius had established that the sourness of soils was due to their acidity and that the sweetening effect came about because limey materials were alkaline, but he then faltered by suggesting that somehow this involved dissolving the 'fattiness' of the soil. He did, though, correctly point out that sweetening the soil is temporary, thus explaining a proverb of the time: 'Liming soil makes fathers rich but sons poor.'

It was in around 1840 at Rothamsted Manor in Hertfordshire, England, that scientific experiments were first carried out to find exactly how this lime effect worked. There the celebrated Broadbalk Field yielded the first understanding of how soil acidity, which depends on the lime content, affects nutrient availability to plants. (One visiting American later remarked that 'we have learnt more about agriculture from this field than from any other patch in the world'.) Today, a farmer can simply tap the 'Rothline App' on their cell phone to find instantly how much lime to add to improve their soil's nutrition.

Many agriculturalists, however, still rely on the standard chart that cleverly depicts how nutrient availability comes and goes with differing soil pH values. It was invented by Nicholas Pettinger of the Virginia Agricultural Experiment Station, who died tragically young, with his contribution now rather unacknowledged. Early in 1935, besides being busy with such things as university lectures, talks to the local Farmers' Institutes and chairing committees, Pettinger was busy preparing his 'Useful Chart For Teaching The Relation of Soil Reaction to The Availability of Plant Nutrients to Crops'. This was the origin of today's much-used table. Tragically, however, while developing his scheme he contracted a serious blood disease and a mere six months later, aged just 35, he died. Albeit unsung, his efforts live on: look in any popular account of soil fertility today and you will likely see some version of Pettinger's pH calculations, an example of which can be seen on the next page.

Most bedrock tends to yield rather acid soils, typically with a pH of 5 or 6 at the most. A glance at Pettinger's chart shows that at such values the vine's access to essential nutrients such as magnesium, phosphorus and molybdenum will be curtailed, such that root growth and plant development will suffer. Richly purple and deep emerald vine leaves may look strikingly beautiful but are probably a sign that they have too little magnesium. But soils with a greater lime content will approach an ideal pH of around 6.5, that of many calcareous soils. Here, access to essential nutrients is at its optimum.

Further virtues of calcareous soils come from their calcium interacting with any clay to weaken the bonding of soil particles. This makes the soil open up to give spaces that not only promote warmth, good drainage and root penetration but also enhance the diversity and activity of micro-organisms. While we still don't know to what extent the soil microbiology affects wine flavour, there's no doubt that a healthy, living soil has huge ecological benefits and can indirectly benefit the finished wine.

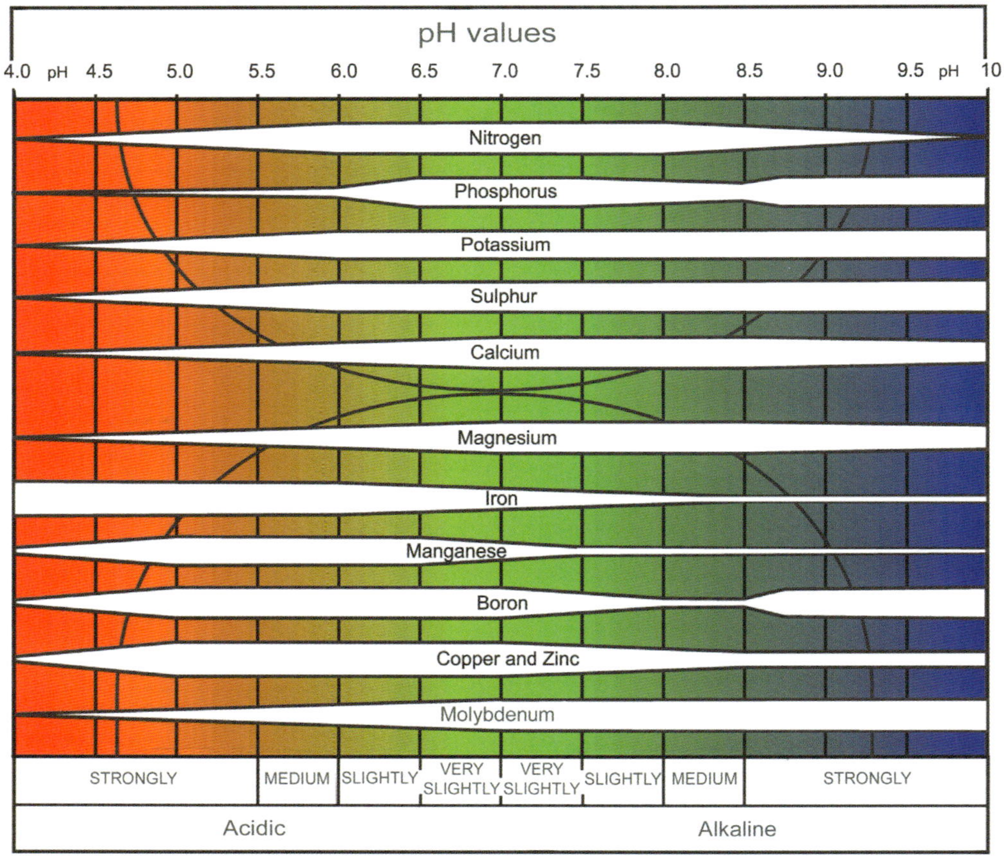

ABOVE: A chart, originally devised by Nicholas Pettinger, showing the relationship between soil pH and nutrient availability.

So in ways such as these, a 'sweet' soil is desirable, and that usually means one with a good lime content. Does that mean, therefore, that a limestone soil is indeed the holy grail? Hmmm. Unfortunately, too much lime can be problematic, and a pure limestone soil consists of little else but lime.

So what's the catch?

Two major challenges can face growers working with limestone soils: the possibility of nutrient deficiencies, and problems arising from soil pathogens, especially phylloxera.

First, nutrition. Purer limestone soils contain less clay, quartz and other silicate minerals: calcite increasingly dominates and as a result pH levels increase. As Pettinger shows, this can introduce nutritional problems. Take boron, for example, with an availability starting to decline at pH 7. It's needed by grapevines in only tiny amounts but even so, boron deficiency is a widespread micronutrient issue. It ruined one of Japan's major wine-producing areas, Yamanashi Prefecture, until it was established that one of the causes was over-liming of the soils. Some of the soils at Montalcino in Italy reach pH values approaching 8 and as a result most of the soils are lacking or severely lacking in accessible boron. It's a similar story with zinc: more than half of the Montalcino soils lack available zinc.

One of the results of boron and zinc deficiencies is shrivelling of the vine leaves. In South Africa's Cape Province, even though zinc is mined in the northwest of the region, in the wine-producing southwest it's sparse in the soils: as a result, the growers there talk about '*kleinblaar*' – little-leaf disease. Another symptom is uneven fruit set, known in France as *millerandage* and in English, picturesquely, as 'hens and chickens' or 'pumpkins and peas'. Not good news for the harvest.

But the big problem is with iron. The difficulty here is not only that limestone soils contain less iron than those derived from, say, basalt or schist, but that the element is usually locked into insoluble compounds that roots cannot absorb. The problem worsens with higher values of soil pH, and a one-unit increase in pH decreases the solubility of iron a thousandfold. The ensuing lack of iron makes it difficult for the leaves to produce chlorophyll, and as a result they turn yellow. It's called iron chlorosis; if you're a gardener you may have seen it on your roses, or tomato and pepper plants.

Or you may have seen yellowing, chlorotic leaves in vineyards, as it's not uncommon in areas with limestone soils. In the Corbières district of France's

Languedoc, two or three rows of vines can look distinctly yellow among the rich green foliage of neighbouring rows. That's because the intricate geology of the district can give very thin strips of high pH limestone soils, even within a single vineyard.

Iron deficiency has been a long-standing problem in the chalky soils of the Champagne region. For centuries the ground there has been supplemented with coaly materials containing iron: coal waste was brought from the mines around Lille, not far away; the low-quality lignite that occurs around Reims was dug out of numerous small pits for this purpose. Several still operate. Right up until the end of last century, in a further attempt to 'improve' the soils, municipal waste (the infamous '*boues de ville*'), containing things like broken glass, batteries and plastic bags, was spread on the ground. The soils of Champagne might attract the holy grail sobriquet but, writing as recently as 1984, one French authority remarked on their 'foul odour' and the 'uninviting aspect to the vineyards, disconcerting for the visitor who too often sees the spaces strewn with plastic debris'.

These nutritional problems might seem surprising seeing as vinifera vines had largely originated and for millennia flourished on calcareous soils. In fact, as well as growing perfectly well on most other kinds of soil, for limestones the species has a cunning mechanism to combat chlorosis: the roots are able to pump out hydrogen and organic acids into the immediately adjacent soil in order to increase its acidity and thus help the iron become soluble and available for uptake. What, then, has gone wrong? Why are there now nutritional problems with limestone? The answer lies with the dreaded phylloxera.

The root of the matter

The watery empire of the goddess Tethys included the great and mysterious sea that stretched out westwards from the Mediterranean: the Sea of Atlas. Or as we know it, the Atlantic Ocean. Geologically, the North Atlantic began to form about 130 million years ago, when the continent that until then had contained Eurasia and North America began to split apart. As a result, the existing terrestrial life forms now found themselves on diverging land masses, on opposing sides of the infant ocean. They proceeded to evolve separately, and this included the ancestors of the plants we now know as the genus *Vitis*, the grapevine.

As we have already seen, in Eurasia *Vitis vinifera* evolved largely on the calcareous deposits from the Tethyan sea floor. But on the American side of the new ocean various different species arose, such as *Vitis labrusca*, which developed on the distinctly acidic gneiss and granite soils of what is now New England. Alongside it, a root-nibbling aphid was also evolving and the grapevines naturally adapted to its presence.

The history of the importation of American phylloxera into 18th-century Europe and the ensuing devastation of defenceless European vines is well documented, including how the most effective solution turned out to be grafting *vinifera*, with its superior tasting fruit, onto the phylloxera-resistant rootstocks of North American vines. Less recounted, though, is the role that limestone soils played in the story. Here's an outline.

Early French attempts at grafting used *Vitis riparia* as, being the most widespread North American wild vine, it clearly had good phylloxera resistance. But as its name indicates – *riparia* meaning to do with rivers – it flourishes on moist, fertile riverbanks and so this presented a problem in the dry, calcareous soils of Europe. So rootstocks of *Vitis rupestris* were tried and – *rupestris* meaning rock living – these fared better in stony soils, but again not if they were calcareous. Might there be an American wild vine living happily in alkaline, calcareous soils? The beleaguered *viticulteurs* were becoming desperate and urged government action.

So it was that in 1887 a young botanist, Pierre Viala, was sent to the United States to search for a grapevine that was tolerant of limestone soils. He visited a host of vineyards in the eastern states but with little success, largely because there was little limestone exposed at the land surface. And wherever he did find some, any local vines were invariably struggling. 'Not one of the varieties of the North and the East has value for calcareous and marly soils,' he concluded.

Viala was sent extra funding in order to continue further west, across 'Indian Territory', and eventually all the way to the West Coast. But there he found imported European vines already decimated by phylloxera – and no limestone. It must have been demoralizing for the young man, knowing that so much rested on his shoulders, yet things kept going so wrong. His diary one day records: 'Despair and woe! Now my little suitcase with my most precious possessions has vanished: all my collected notes since I've been here... I cried like a child.' (It was later recovered.)

Then came a change in fortune. After a tip-off, Viala travelled to the little Texas town of Denison, north of Dallas, right up by the Oklahoma state line, to meet Thomas Volney Munson, an indefatigable cataloguer of American vines. (The two immediately hit it off: later Munson named one of his daughters Viala!) Munson not only understood vines but he knew their habitats and, crucially, the soils they grew in. And yes, he knew exactly where vines were thriving on rocky limestone.

Thus Viala rode down to Texas Hill Country, to a place just west of Belton called Dog Ridge. There, he found limestone soils remarkably similar to those in France and with vines growing contentedly in them. He swiftly identified the particular species that Munson had recommended, *Vitis berlandieri*, and soon 15 wagon-loads of cuttings were being taken away and loaded onto three ships bound for France. Salvation for the *viticulteurs* was on its way!

Viala's return, however, wasn't wholly triumphant. Among other things, back in France his cuttings proved reluctant to take root, such that scientists had to resort to hoping that selective crossings might produce some viable offspring. And eventually they did. A cross between *Vitis berlandieri* and Chasselas (*Vitis vinifera*) yielded the famous rootstock 41B. It rescued the particularly badly stricken Charente vineyards and is still used for more than 80 percent of the vines in the Champagne region. But while the *vinifera* component of such hybrid rootstocks gives the desired superior grapes and the *berlandieri* helps fend off phylloxera, there are still the chlorosis challenges. Usually, components of other North American rootstocks with better rooting tendencies are needed in the crossings and they dislike high pH limestone soils.

The shortcomings work in reverse: early vineyards in the USA clustered along Lake Erie and the southern Finger Lakes where the indigenous vines were well suited to the acid gravels, sands and shales, but produced mediocre wine. Growers avoided the more northerly areas towards the Niagara limestone escarpment where the local vines struggled. Up there, though, introduced *vinifera* was quite at home nutritionally, but on its own roots defenceless against rampant phylloxera.

Different, yes – but special?

So is limestone special for wine? Although the rock has never had a monopoly on superlative wines, arguably there was justification for the claim in pre-phylloxera days when *vinifera* grew happily in limestone soils on its own roots. There are plenty of modern commentators who believe that those pre-phylloxera vines produced wines unmatched today. Nowadays, however, it's a risky proposition to use ungrafted vines, even in regions that so far have avoided phylloxera, because of the ever-present and perhaps increasing chance of infestation by a whole array of soil pathogens. As a result, most modern growers resort to grafting, which involves the careful selection, after weighing various pros and cons, of an appropriate rootstock. And consequently having to accept some degree of compromise.

It's worth pointing out that limestone soils seem particularly attractive to phylloxera, exactly as the French experience illustrates. In volcanic and sandy soils, however, the insect struggles: the few surviving pockets of pre-phylloxera vines in France are almost all on distinctly sandy soils.

Many of the physical attributes of limestone can be emulated by other kinds of rock. Plenty of soils are free draining and, usually through clay being present, can offer water storage. I mentioned that most limestone wine cellars aren't natural caves but have been fashioned from disused quarry workings, so, similarly, cellars can be excavated in many rock types. They range from soft materials such as loess, as in Kamptal, Austria, to tough granite, as in South Africa's Western Cape. In Hungary, UNESCO's huge Rákóczi cellar in Sárospatak and four-storey 15-kilometre Ungvári wine cellar in Sátoraljaújhely are carved from volcanic tuff.

A limestone finesse

One supposed attraction of limestone soils is the distinctive quality they give to wine character (apparently irrespective of the vine rootstock), though there is much inconsistency in the claims. Most commonly it's expressed with words such as 'liveliness', 'edge', 'nervousness' and 'finesse', perhaps in line with the notion that the alkaline, high pH soils produce low pH wines. The thinking possibly stems from the revered and influential wines of northern France, especially from Burgundy and Champagne and their calcareous soils, though even in the Côte d'Or for some writers the limestone soils lead to 'big, strapping wines' and 'unmatched opulence'.

Many wines from northerly countries such as Germany, Austria and Hungary attract similar descriptions of 'finesse' but are from acidic, not calcareous soils. Conversely, wines from limestone in warmer climates tend not to be noticeably acidic – like Chardonnays from Lebanon's Beka'a Valley – as with most red wines.

Presumably the greater acid levels from these northerly areas have a lot to do with cool climate, but designing a scientific experiment to isolate and assess the contribution of the soil is a formidable challenge. Yet in a recent Dutch experiment the obstacles were circumvented. Grapes from different cultivars were grown similarly, but each in a different soil, and were then uniformly vinified. Trained tasters then assessed the wines but were unable to detect any significant differences among them irrespective of the soils. In an extension of the experiment, the tasters did detect variations when the wines were made using different yeasts, but the soils, which included limestone, appeared to be playing no role in the taste.

It's instructive to look at areas where limestone occurs in close association with other rock types. The Leithaberg DAC in Austria is dominated by two soils, limestone and slate, but the official regional specification makes no differentiation between them. It's similar at Saint-Chinian in southern France, where the appellation covers both the limestone soils in the south of the region and the schist in the hillier north. Interestingly, and in contrast with northern France, commentators find fuller, more powerful and fatter wines coming from the limestone soils whereas it's the schist soils that yield wines with more acidity, tension, elegance and perfume. The schistose areas are higher in altitude so a cooler climate may be involved, but then in Bourgueil, on the Loire, light, fruity, 'thirst-quenching' wines come from the gravel down by the river and 'richer, well-built, age-worthy wines' from the higher limestone bluffs.

It would seem, then, that limestone doesn't confer a special character on wine, at least in a consistent way. And anyway, does the idea that limestone soils in a vineyard are something of a holy grail – if that implies a kind of ideal – square with the viticultural challenges they present, at any rate where the parent rock is fairly pure?

Near where I am writing this in mid-Wales is the old manor house of Nanteos, which until recently kept a wooden object believed by some Christians to be the Holy Grail. Supposedly, following the Crucifixion it had been brought by Joseph of Arimathea to Glastonbury Abbey in England, then at the time of the Dissolution of the Monasteries spirited to Nanteos for safe keeping. There, over the ensuing centuries, it was held sacred, bringing miracles to all who beheld it, as befitted a fragment of the True Cross. Until, that is, radiocarbon dating spoiled the whole story. So perhaps the search for the holy grail still goes on…

RIGHT: Dawn on the limestone cliffs of Mont Granier, from the Apremont vineyard, Savoie, France.

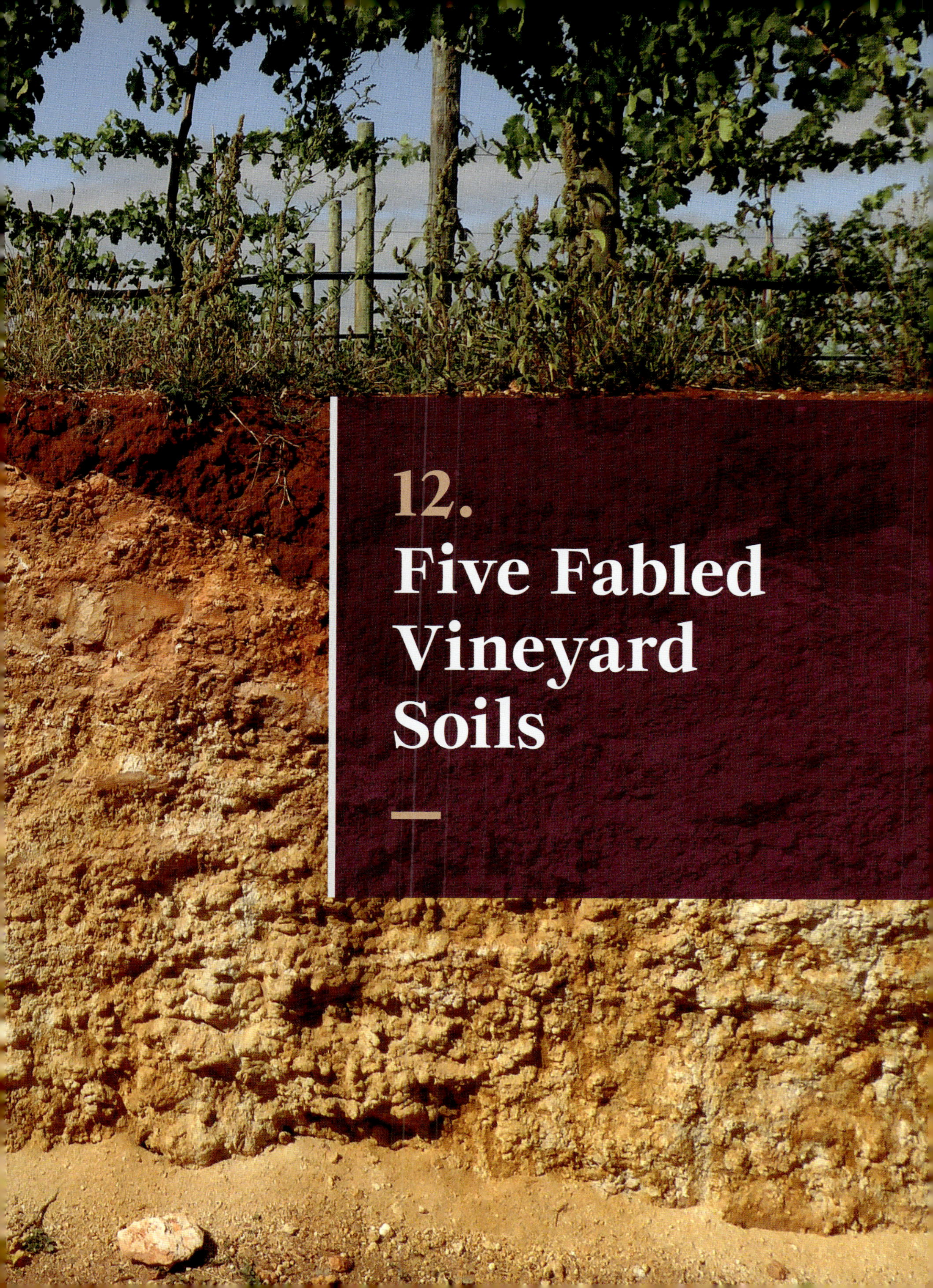

12.
Five Fabled Vineyard Soils

Five Fabled Vineyard Soils

When young William Gimblett sailed out of London in 1878 headed for a new life in New Zealand, he must have had little idea of what lay ahead of him. He is scarcely likely to have imagined that 30 years later he would be raising sheep and acclaimed for his precocious 'Gimblett lambs'. That came about because after settling in the North Island's Hawke's Bay he purchased some land just inland from Napier, strikingly stony but nicely sheltered, and its relative warmth meant his lambs reached maturity considerably earlier than those of his neighbours. And he is extremely unlikely to have imagined that 100 years after that the land would be bearing grapevines, with those self-same warm, gravelly soils called after him and the Gimblett name known to wine enthusiasts the world over.

All the world's grapevines are, of course, growing in soils of one kind or another, which all provide stable anchoring along with water and nutrients. But in just a few places they have acquired a designation of their own: names that are treasured by wine lovers. So, we have *galestro* and *llicorella*, Willakenzie and *gore*, *roten schiefer* and *galets roulés*: names that probably mean little to most people but which to wine enthusiasts prompt images of special places and wines of distinction. In this article I will explore the stories and the science behind five such named soils, examples which have become legendary in the world of wine.

The Kimmeridgian of Chablis

Where is the largest onshore oilfield in Western Europe? It's in the southwestern UK, and is certainly no new-fangled development. For centuries the folks there

PREVIOUS PAGES: The terra rossa of Coonawarra, showing its 'wavy' but sharp junction with the underlying white (though here slightly iron-stained) limestone. DiGorgio winery, Australia.

knew that some of the local rocks could burn, and from them such things as varnish, grease and paraffin could be manufactured; a 'nodding donkey' outside one of the villages has been pumping oil for more than 60 years. The name of the village? Kimmeridge.

The word is legendary to wine lovers: the Kimmeridgian of Chablis is perhaps the single most famous vineyard soil in the world, and the best loved. Perhaps also it is the most misunderstood. Why so? Well, to start with, my point in mentioning oilfields is to underline that in geology 'Kimmeridgian' denotes a particular period of time and not a kind of rock or soil: there are no oily rocks around the village of Chablis. But the limestones and marls of Chablis formed at the same time as the rocks at Kimmeridge, so both are referred to as Kimmeridgian *in age*.

Similarly, there are rocks at Kirkland Lake, just west of the Quebec–Ontario province line in Canada, that formed deep within the Earth, so deep that they experienced pressures sufficient for diamonds to form in them. They are Kimmeridgian. In northern Turkey there are Kimmeridgian granites, and very different kinds of igneous rock, though also Kimmeridgian, occur in the Coast Ranges of western California. In other words, describing the vineyard soils of Chablis as Kimmeridgian carries little meaning without prior knowledge of what kinds of materials they are.

But (and putting aside the problem of explaining the popular belief that the particular nature of Chablis wines is somehow due, notwithstanding the unique climate and physical setting of the Chablis slopes, to the vineyard soil) there are further misunderstandings about the term.

Wine writers seem to assume that in science the term 'Kimmeridgian' is neatly defined; in fact, geologists have argued about exactly what it means for more than 200 years, sometimes decidedly vituperatively. Some have recommended that the term should be discarded altogether. Finally, however, as recently as 2021, an international scientific agreement was reached on what Kimmeridgian should actually mean. Even so, one continuing problem in practice is that the now-accepted definition is based on certain fossils (very particular kinds of ammonites) that aren't present in many rocks probably of Kimmeridgian age, and this is the case at Chablis.

In any case, for us in the wine world there are two problems that are more fundamental than these technical matters. First, grapevines simply don't care anyway, so to speak, about the geological age of the vineyard rocks. Wine

promoters and commentators love to mention the names of geological periods and often imply that the older the age of the vineyard rocks somehow the more profound and complex the wine. Really, however, it's the physical and chemical properties that are relevant to the vine, not the geological age, no matter how old.

Second, all this business of geological age is referring to the bedrock, the intact substrate of the vineyard. It may have fissures into which deeper vine roots probe for supplementary water, but otherwise it is the overlying, humus-bearing, loose material – the soil – in which most of the vine roots grow and obtain water and nutrition. And almost invariably, *the soil is vastly younger than*

ABOVE: The white, marly soils of Chablis, derived from bedrock of Kimmeridgian age; here at the Grand Cru Les Clos vineyard, overlooking the village of Chablis, Yonne, France.

the bedrock (*see* Chapter 10). In most of the world's temperate zones, certainly in the glaciated northern hemisphere, the soils are no older than a few thousand years and are still forming. In other words, the term 'Kimmeridgian' isn't really referring to the soil at all.

The Kimmeridgian bedrock at Chablis lies beneath the middle parts of the vine-covered slopes, with younger strata capping the hills and older ones at the base. The soil, though, isn't fixed; through time it inevitably creeps downslope. So for centuries growers have had to cart the jumbled debris accumulated at the foot of the slopes back up to where the vines are growing; consequently the mid-slope viticultural soils are rather mixed.

A Google search linking Kimmeridgian and Chablis yields thousands of results, nearly all enthusing about the importance of the connection between the two. Few, though, suggest there are any difficulties.

The albarizas of sherry country

'Architectural marvels', diatoms and radiolaria have been called. The filigreed arrangement of their glassy white shells has inspired structures such as geodesic domes as well as visionary furniture. Both kinds of creatures are extremely tiny – they're varieties of microscopic oceanic plankton – and they are both made of silica (silicon and oxygen). There are other similarly intricate planktonic organisms such as foraminifera, but with shells made of white calcium carbonate. All of them float in the world's oceans in prodigious numbers and sometimes, if conditions are right, one of the varieties can flourish in even more extraordinary numbers, with their fragmented dead shells dominating the ocean-floor sediments.

Exactly that happened around 20 million years ago, in the seaway that at that time was connecting the Mediterranean to the Atlantic. (The sea floor eventually became uplifted to make land, the region of Spain we now know as Andalucía.) From time to time sands or muds would sweep in from the nearby land, and although diatoms flourished throughout, populations of the other plankton fluctuated as conditions in the sea water varied. And as we will see shortly, today all this has influence on the drink we call sherry.

So, back in time a series of sediments was built up on the sea floor, eventually to be hardened into rocks with the planktonic organisms preserved as microfossils. Today those rocks, essentially marls, form the Andalusian landscape and are weathering into soils. They are known there as *moronitas* and all are rich in diatoms.

In the Jerez-Sanlúcar sherry district they have vernacular names such as *arenas* for those that are sandier and *barros* for those richer in clay. And, for those particularly rich in calcareous plankton – all pale coloured but in places startlingly white – albarizas. These well-drained, low-fertility soils yield the finest grapes for sherry, not least because they tend to occupy the tops of the area's rolling hills and hence benefit crucially from cooling sea breezes. But also because the fossilized hollow shells of the plankton together with the little spaces between them give the soils a loose structure that allows incursion of extensive root systems and an excellent capacity for storing rainwater – invaluable during the arid growing seasons of Andalucía. A piece of albariza, surprisingly light because of all the little holes, can absorb water up to a third of its volume, rather like a sponge. (These white, powdery albarizas are often called chalk though, geologically speaking, they contain insufficient amounts of the microfossil [*coccolithophores*] that defines true chalk and they formed about 70 million years after the definitive chalk of England and northern France.)

With time, the region's growers learned of sub-types of the albariza soils and demarcated the different wines they yielded. During the last century, however, such nuances became lost as the market for sherry dwindled and wine-store shelves became dominated by heavily promoted brands. Wine distinctions deriving from their terroirs vanished.

Today a new generation of producers is striving to restore the status of sherry as a fine wine and to resuscitate the characteristics of different vineyards. The albarizas terroir scheme devised by Isidro García del Barrio Ambrosy, for instance, has seven different sub-classes, based on such things as altitude and slope as well as the soils.

Right now, there are around 20 producers selling wines emphasizing their derivation from different types of albariza. Examples include tosca de barajuelas, the soil richest in diatoms and hence the lightest and most porous, with a leafy, laminated structure that encourages the roots to grow sideways. Tosca cerrada is more calcareous, with the lowest proportion of siliceous diatoms, soft when wet but becoming baked when dry. The soft, foraminifera-rich Tosca lentejuelas (literally 'sequins') encourages deeper roots, which is said to give lighter, more elegant wines.

Almost certainly the contribution of these soils to the differences in the wines arises primarily from fine variations in their water-holding capacities,

OPPOSITE: Ripe Palomino grapes in Andalucía, Spain, destined for sherry and growing on albariza – white, marly soils rich in calcareous plankton.

due to various proportions and compactions of sand grains, clay and the different microfossils. But there may be additional effects. Some growers say that the thicker grape skins produced on softer albarizas such as *barajuelas* bear a lesser proportion of wild, indigenous yeasts and that the soils may even influence the kind of yeast. Some say that softer albariza soils encourage the vigorous beticus strain whereas wines from denser soils have a higher chance of developing the gentler montuliensis. These are both film (surface) yeasts and so are intrinsic to the flor that characterizes sherry.

Perhaps, then, sherry is on the threshold of a new terroir-driven era, with the albarizas regaining their celebrated status and familiarity. (The word is already appearing, through the sherry casks used for ageing, on the labels of some whiskies!) And central to it all, down there in the Andalusian soils, are those little architectural marvels.

Napa's Rutherford Dust

It's a catchy phrase, Rutherford Dust. Growers in California's Napa Valley love to use it and, amongst other things, it's appeared in poetry, as the trademark of a bottled water, and as the title of pieces of music. But in the wine world it has mythical status, and some would say literally so. For whether it actually refers to a soil is much debated: what the term really means is far from clear.

There seems to be no record of the expression's origin; how and when it was first used is clothed in anecdote. Some say it's due to the great UC Davis academic Maynard Amerine. That's not hard to imagine, seeing as Amerine was so fond of sayings: 'Wine quality is easier to detect than define'; 'Drink wines, not labels'; 'Wine is a chemical symphony'; and so on).

But most point to the much-quoted saying by André Tchelistcheff: 'It takes Rutherford Dust to make great Cabernet.' At face value this seems a pretty clear reference to soil, but Tchelistcheff, being winemaker at Rutherford's Beaulieu Vineyards, would have known very well that the soil there is a far cry from being dusty. Dust is fine stuff, carried by the air. The Rutherford AVA stretches eastwards from alluvial gravels across the valley of the Napa River, which over time has deposited spreads of loams, sands and silts. Nothing as fine as dust, and none of it ever airborne.

The western parts of the AVA flank the Mayacamas Mountains and are slightly higher than the valley floor, on a landform colloquially called in this part of the

world a 'bench'. The Rutherford Bench is home to such revered wineries as Inglenook, Niebaum-Coppola and Far Niente, as well as Beaulieu, and consequently many a wine-loving tourist wants to see the celebrated feature. The relative elevation is only slight, however, and the slope onto it barely noticeable, say, as you drive along Highway 19. So one or two institutions here have thoughtfully catered for sightseers by providing a fittingly named wooden bench, for sitting and selfies.

The real Rutherford Bench is an alluvial fan, formed by streams emerging from the Mayacamas mountains and immediately dropping their coarsest sediment. Less coarse material is carried further eastwards, down to the middle parts of the apron. The finest sediment forms the outer fringes, where it interfingers with the alluvium of the Napa River itself. The soils are of suitably limited fertility and superbly drained – some vine roots probe 16

ABOVE: Rutherford Dust is feted in California's Napa Valley, but what the term actually means is much debated. The soils at Rutherford are on an alluvial fan and so are far from being 'dusty'.

metres (50 feet) or more to find their water. In reality it's all more complex than this, not least because through time the streams switch their routes and braid across the fan, which is why the soil properties vary so intricately.

Why, then, did Tchelistcheff (or whoever) refer to these coarse soils as dust? Certainly, there are those who believe the term does refer to the soil: 'Rutherford dust is a mixture of gravel, sand, loam and volcanic deposits that permeates the Rutherford bench plateau' (all *sic*); 'we walked between the vine rows, kicking up the famous Rutherford dust'. Perhaps Tchelistcheff really meant 'dirt', in the colloquial American sense?

More commonly, the expression is taken to refer not just to the soils but to the overall characteristics of the area: Rutherford Dust 'essentially refers to the unique terroir particularly with respect to Cabernet Sauvignon'. The Rutherford bench experiences Napa's warmth but is nicely tempered by maritime air drifting northwards from San Francisco Bay; it receives relatively more sun exposure through being located at the Valley's widest point yet has a significant diurnal temperature variation. These special combinations lead to wines of special character and many believe that the true meaning of the phrase lies in the wine itself.

But even this is unclear. Some think of aroma: 'an intriguing aromatic element often referred to as Rutherford Dust'; 'the signature Rutherford dust smell'; whereas others of taste: 'a mysterious, spicy element known as "Rutherford dust"'; 'the wines are very Rutherfordesque. You can taste the dust'.

Mouthfeel, however, is most commonly mentioned: 'a nuance attached to the back of the mouth'; 'Tchelistcheff was talking about granular or "dusty" tannins that couple with cocoa powder'. Tannins are frequently mentioned in tasting notes on Rutherford wines, and research has shown that a restricted water provision, exactly as these bench soils provide, is an important factor in developing grape tannins.

Given the experience and insight of such people as Amerine and Tchelistcheff, it seems likely that something like this was in mind when the term was coined. That's the view of the association of local wineries, the Rutherford Dust Society: the word 'dust' 'evokes in four simple letters a specific place and a complex set of flavours'. As Andy Beckstoffer, who worked with Tchelistcheff, put it: 'When he said "the wines must have Rutherford dust in them" he did not mean they had to taste of dust. André meant they needed to taste like they came from Rutherford's vineyards.'

Even so, there are those who see the phrase as purely allegorical, referring figuratively to a 'magic dust', sprinkled over this charmed area. Or, perhaps with a touch of cynicism, a 'gold dust', reflecting the dizzying prices that some of the area's wines can command. (Tchelistcheff once said that 'Money was the dust of life'...)

Yes, Rutherford Dust is a catchy phrase but its meaning is imprecise, apparently lying somewhere in a blurry matrix of interpretations. Perhaps – not unlike some other words in the wine lexicon, and to paraphrase the words of Humpty Dumpty – it means what you want it to mean.

The terra rossa of Coonawarra

Picture the scene: the Australian outback. A vast horizon shimmering beneath an azure sky, bushy acacias and perhaps eucalyptus, maybe even a kangaroo or two. And, of course, the red ground. In places dazzlingly red. Much of interior Australia has a distinctly ruddy look, and it's essentially due to iron.

Warm regions across the world tend to have soils that are more or less red, depending on the proportions of certain iron minerals. Most widespread is the rusty-orange, ocherous goethite, named after the German polymath and avid mineral collector. Hematite, as its name suggests, is the colour of blood, and it's this that makes some soils so vividly red. That effect is seen widely in Mediterranean areas and there the soil has long been venerated. Since ancient times it's been used as a pigment in sacred tombs, and in certain religions it's seen as the very dust from which God created man. (In biblical Hebrew the soil is called *Adamah*, *dam* meaning blood and *Adam* meaning red).

In recent centuries, the reddest soils have become known by the Italian name terra rossa and science has long been grappling with understanding them. Exactly how red do the soils have to be to justify the name? Hence, scientists juggle with the hues (specific colours), values (lightness and darkness) and chromas (colour intensities) to try to agree on limits of appropriate redness.

How do they form? There is general agreement that true terra rossa forms only on limestone and usually involves the slow dissolving away of calcium carbonate to leave an insoluble residue of clays stained by hematite. But in some places, rather than being a residue, the limestone has been slowly replaced by iron and clays in situ, by capillary waters rising through the rock. And elsewhere, chemical analyses show a substantial input of red material that must have originated somewhere else, such as the wind-blown Sahara dust in the terra rossa of Croatia's Istrian peninsula.

There's also a consensus that terra rossa formation needs warm summers but wetter winters, and well-aerated, oxygenized conditions so that the insoluble oxidized form of iron – ferric oxide, the mineral hematite – forms preferentially over hydrous goethite and other forms. Just such conditions arose in South Australia, in the region now called the Limestone Coast. And as in Mediterranean areas, the resulting terra rossa soils have proved conducive for growing quality grapes, as in today's noted wine regions of Robe, Mount Benson, Wrattonbully and Padthaway. And, of course, Coonawarra.

The Scotsman John Riddoch first succeeded with Coonawarra wine, such that by 1896 he was able to build a fine sandstone winery, with a distinctive triple-gabled frontage. However, for various reasons his initial success gradually faded, though the locals remained convinced of the area's brilliant potential. Eventually, after decades of neglect, in 1951 the Wynn family took the bold step of buying Riddoch's old winery and its surrounding vineyards with a view to producing world-class wines. It worked. Just two years later 'Wynns Coonawarra Estate Claret' was being exported to England – transported on the royal yacht *Britannia*, no less. Today the Wynns label with its triple-gabled insignia is one of the wine world's most iconic.

Nowadays there is little of Coonawarra's narrow strip of terra rossa that is not carpeted by thriving vineyards. Some of them produce grapes under contract, others supply the three dozen or so cellar doors. Several of the latter are so proud of their soils that they maintain a trench in order that visitors can clamber down to see how the vine roots probe through the red soil to try to penetrate the hard but fractured limestone below (*see* pages 192–93).

Two things are immediately noticeable in the trench walls. One is the sharpness, the abruptness, of the change from the terra rossa to the underlying white limestone. Second is the 'waviness' of the line that separates the two, representing in three dimensions an irregular, uneven boundary, just the kind of surface that limestone forms when exposed to the atmosphere: the fissures and pits of karst.

Plenty of wine writings still refer to Coonawarra's terra rossa being an in situ residue but it's more complicated than that. During the last half a million years or so, the Limestone Coast region consisted of a series of coastal reefs and dunes, changing position and height as sea levels fluctuated. Accumulations of wind-blown calcareous clayey silt, sitting on a porous limestone bedrock and filling its surface hollows, produced the higher ridges. The Coonawarra strip is just such a rise, merely metres higher than the surrounding damp land but sufficient to give a drainage and aeration that leaches the calcium from the wind-blown deposits to leave insoluble clays reddened by hematite.

Today the resulting terra rossa provides a restrained fertility ideal for grapevines together with some water storage, crucially enhanced by the underlying fissured limestone. Moreover, the rise in elevation, albeit slight, raises air dryness and exposure to sunlight. Consequently, the Cabernet grapes that Coonawarra is noted for are smaller and more intensely coloured than those from the adjacent, moist soils, and their phenols ripen well to give the area's characteristic blackcurrant and minty wines. The wine colour, incidentally, is largely due to anthocyanins, the red pigments in the grape skins and, despite what some writers assert, it has no connection with the redness of the soil.

Coonawarra is hardly a tourist mecca. It's located away from the region's main highways and is a strip of virtually flat vineyards with mostly new winery buildings; the settlement itself comprises little more than a shop, a restaurant and a few cottages. Yet to watch its terra rossa glowing under the setting of a flaming Australian sun is a wondrous sight, worthy of one the world's most famous vineyard soils.

The Gimblett Gravels of Hawke's Bay

Much of New Zealand's North Island 150 years ago belonged to the Māori, but European settlers were arriving thick and fast, seeking to claim the most fertile tracts. The Hawke's Bay area was attractive: three major rivers were bringing in spreads of fertile alluvium, its climate was 'proverbially mild and healthy' and the Crown was negotiating the purchase of blocks of land. But there was a snag for the new settlers, especially farmers. The threat of flooding.

Māori lore told of devastations resulting from periods of intense rain and swollen rivers, every 40 years or so. Thus it was in 1867, when in 10 days more than 38 centimetres (15 inches) of rain fell and the rivers broke their banks: 'The whole country from the hills in the extreme distance to the town of Napier was one vast sheet of water.' And when the floodwaters eventually subsided, the Ngaruroro River had changed its position and by five years later had completely deserted its old course. The pebbles and boulders that once formed the riverbed were left high and dry.

This new, stony ground was of little interest to the established farmers, but some settlers tried to see if anything could be done with it. There were efforts at sheep farming, and so it was that William Gimblett – he of the introduction to this chapter – purchased some land and tried his hand at it. But his initial success with early lambs was not to last; the barren shingle was simply unable to sustain grazing pasture year after year. He had to move on. He vacated the land and the track he had used for access was gated off, though still vaguely referred to as Gimblett's.

That riverbed land seemed pretty useless. One writer remarked that even rabbits wouldn't venture onto it – without taking a packed lunch. However, in 1981 another emigrant Englishman saw a new possibility. Chris ('C J') Pask had a few vineyard plots in coastal Hawke's Bay and with his little Auster aircraft ran an aerial fertilization business. His plots could produce decent Merlots but the great prize of the time, Cabernet Sauvignon, was in most years too unripe and herbaceous. Cabernet was deemed 'a difficult cultivar'.

One day, however, while flying over the dry pebbles of that old riverbed, C J was reminded of the gravelly soils he had seen in the Médoc and wondered if his Cabernets would have a better chance of ripening on that shingle – with a bit of his aerial top dressing. The inland areas were a few degrees warmer than near the coast and new measurements were showing the soil temperatures to be warmer still. That looked promising.

Several other growers took notice and planted a few vines. And to some surprise, when the wines were ready they attracted considerable acclaim; what followed is now the subject of textbooks and scholarly articles on wine marketing. The growers organized themselves into a formal association and, instead of the usual approach of seeking an official appellation or protected geographical identity (a device that had never been embraced much by New Zealand producers anyway), applied for a registered trade mark. The group didn't specify techniques or grape varieties but sought to 'protect and develop the reputation of the area' by precisely defining its limits, laying down strict quality requirements for the wines and, of course, inventing a brand name. The area's striking character lay in its river gravels, centered on William Gimblett's old track. So the brand name simply fell out: Gimblett Gravels.

The chief contribution of the gravelly soils is their free drainage. This results in a low soil fertility that restricts both the vine canopy and the sward between vine rows, enabling the bare stones to absorb the day's heat for re-radiation at night. There is just enough topsoil to provide the humus-borne essential nitrogen, phosphorus and sulphur but all the other nutrients have to come from irrigation. Hawke's Bay waters come largely from the inland mountains, rising among the deposits of the great Taupo volcano and passing eastwards through a variety

of sedimentary rocks such as the very impure sandstone called greywacke; consequently they easily have sufficient dissolved nutrients.

Food purists may scorn tomatoes and salad crops produced hydroponically, but that is essentially the way these premium grapes are being grown. Viticulture here is impossible without irrigation; vines begin to fail if rodents chew through the supply pipes! So it's an interesting observation that although wines from the Gimblett Gravels are often cited as examples – outstanding examples – of terroir-driven wines, the grapevines' vital water and nutrient supply comes from far away.

Today, the area's Cabernets are of international stature and other red grapes such as Syrah, Tempranillo and Malbec are fast gaining in repute; competition for space within this restricted acreage is fierce. A couple of statistics sum up the explosive success of the Gimblett Gravels. In the early 1980s a hectare of land here was changing hands for less than $10,000; in 2020 the rate was over $150,000. Provided, that is, it was within the area entitled to the Gimblett name. William himself would surely be amazed.

ABOVE: Gimblett Gravels at Deerstalkers vineyard, Hawke's Bay, New Zealand. These decidedly stony soils yield some of the country's finest full-bodied reds, such as Cabernets and Syrahs.

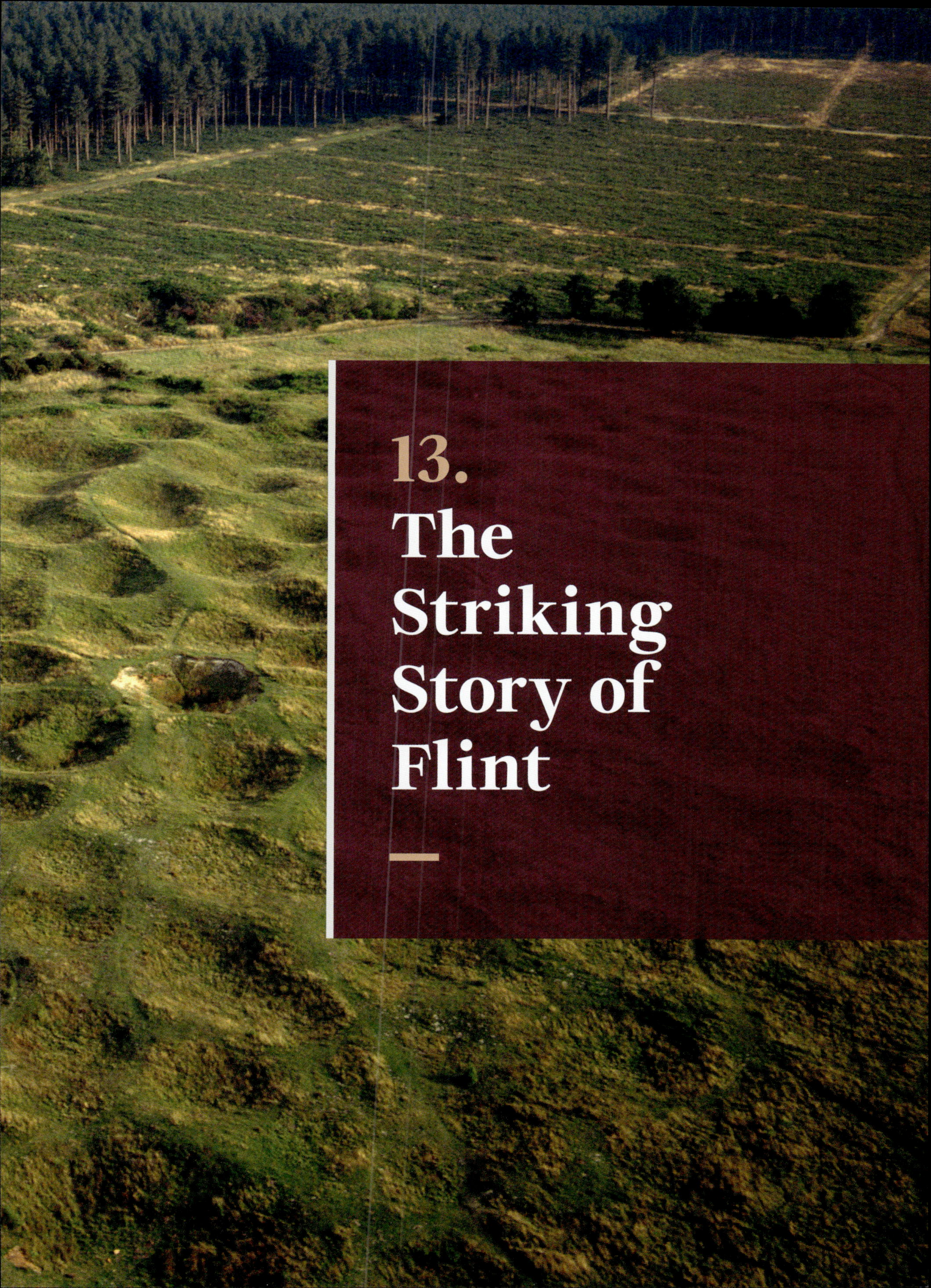

13. The Striking Story of Flint

The Striking Story of Flint

Everyone has heard of flint. To some it means prehistoric tools, to others it's the gizmo that makes sparks in cigarette lighters, and to many farmers it's the damnably hard field stone that wrecks their machinery. But to readers of this book the word almost certainly conjures up certain kinds of wine, Chablis and Sancerre perhaps. 'A quintessential Sauvignon Blanc with the tell-tale hint of flint': that kind of thing. In fact, words like flint and flinty are these days among the most commonly seen terms in wine descriptions.

It seems people like the word. You can buy a 'flint' guitar and 'flint' shoes (oddly enough from a shop called Margaux). You can buy 'flint' coconut bars and 'flint' bread snacks (caviar-flavoured, no less). It crops up in popular culture – Van Morrison's voice has been likened to flint – and it's a much-used word in literature. Ebenezer Scrooge's heart was 'hard and sharp as flint'. This multitude of allusions presumably stems from the material flint itself and the way we perceive what it is, how it behaves. So what exactly is this potent, charismatic substance?

Flint fundamentals

Flint is made of silica, that is, silicon dioxide. The two elements silicon and oxygen are bonded together in an ordered three-dimensional framework, what in science is called a crystalline arrangement, but each crystal in the flint mass is truly tiny. It's no more than 20 thousandths of a millimetre across, about the thickness of plastic clingwrap. Geologically, it's this extreme fineness that distinguishes flint from silica minerals such as quartz and amethyst, with their sizeable crystals.

It makes the material hard and tough, as implied in the literary allusions mentioned above, and this has practical implications for vineyards. It also

PREVIOUS PAGES: Grimes Graves, Brandon, Norfolk, England. Each of the surface depressions was in prehistoric times a pit (now largely filled in), dug in order to extract flint.

makes flint chemically inert and virtually insoluble, which, amongst other things, means it is tasteless and odourless. Hmmm. This raises tricky questions about what we actually mean by saying a wine tastes or smells 'flinty'. I will be returning to this intriguing issue.

So how does flint come about in nature? Essentially, it forms from the decayed silica skeletons of tiny creatures that inhabit the world's oceans. Favourable conditions for this have arisen a number of times in Earth's history, including the present day, but they hit an optimum around 90 million years ago. The climate then was exceptionally warm, sea levels were high, and considerable parts of the Earth's surface were covered by sunlit shallow seas, rather like the Bahamas today. Calcareous creatures such as corals abounded and produced sea-floor sediment that would go on to make limestone (*see* Chapter 11).

However, from time to time there were also huge numbers of little creatures made of silica, such as diatoms (tiny cells of algae said to be floating in the oceans today in such numbers they account for a quarter of all the photosynthesis on Earth) and the beautiful glass-like architectural marvels called radiolaria that we met in the albarizas of Chapter 12. When these organisms died, their silica skeletons accumulated on the sea bottom and with time turned into a gelatinous mass. Then, upon becoming buried a few metres or so, the material became sufficiently reorganized internally to make the finely crystalline matter we know as flint, forming bands and lumps within the calcareous sediment that was to become limestone.

Important for vineyards, however, is the detritus formed by the weathering of these rocks. The toughness of the flint is likely to preserve it as pebbles, to be redistributed by rivers as gravel. And typically, this is how we see it in vineyards: spread out across the ground.

Where in the world?

Flint is not an abundant mineral but it's widely distributed. It pops up in places from southern Scandinavia to the coast of South Australia. (Where, I read, one grower crushes the flint stones in his vineyard into smaller pieces, believing that this leads to greater flintiness in his wines.)

Curiously, pieces of flint are also found in Australia around Sydney Harbour: curious because there is no natural flint anywhere nearby. So where did they come from? The most likely answer is London! It seems that when convict transport ships were being outfitted ready for the long voyage to the colony, stones – including flint – were taken from the banks of the river Thames to serve as ballast and then dumped on arrival in Sydney.

In North America flint is widespread and of various geological ages. It occurs, for example, in Maine, New York, Missouri and Texas. Flint is the official state gemstone of Ohio. Artefacts made from it, dug from Flint Ridge in eastern Licking and western Muskingum counties, have been found right across the present-day eastern United States and as far west as the Rocky Mountains.

But of those warm Cretaceous seas I mentioned earlier, the most extensive development was the huge, albeit shallow, ocean that spread from what is now the British Isles, across northern and central Europe, to the Middle East, Crimea and Kazakhstan. And that is why, as I discuss in Chapter 11, limestone happens to underlie so many of the classic wine regions of Western Europe.

The conditions for flint formation reached their apogee in and around what is now eastern and southern England and north-central France. It was here that prehistoric humans found the best flint for making tools, with perhaps the most desirable material of all coming from around Grimes Graves in Norfolk, England, and Le Grand-Pressigny in the Indre-et-Loire region of central France. More of these two remarkable places later.

In line with the above, parts of East Anglia have very flinty ground. Already well established there is Norfolk's Flint Vineyard, with its eponymous soils.

ABOVE AND OPPOSITE: Flint tools from Grimes Graves, Brandon, Norfolk, England.

Vineyards on England's South and North Downs, together with those in mid-Dorset up through Wiltshire and northern Hampshire, are flinty: examples are Tinwood, Westwell and Langham Wine Estates, and Danebury vineyard. Wiston and Highdown vineyards on the South Downs are only a few kilometres away from the prehistoric Cissbury flint workings.

In France, parts of the Champagne area, such as the Côte des Blancs and the Côte de Sézanne, are famously flinty. So are parts of the Loire Valley, such as at Vouvray. But probably the most-quoted instances are at Pouilly and Sancerre, home of the often-claimed connection between the soils and a flinty aspect to the wines. But besides the scientific difficulties with that, it's noticeable that most of the sites there aren't particularly flinty. In Sancerre, flint soils occur at Saint-Satur, Ménétréol, Thauvenay, and at the top of the hill of Sancerre town itself; in Pouilly, Saint-Andelain is the best-known site for flint, together with Tracy, where some vines are actually planted in an old flint quarry. Even so, other kinds of soils account for fully 80 percent of Pouilly and Sancerre.

Flint-bearing rocks are found sporadically across Eastern Europe and the Middle East, in some cases coinciding with wine-producing areas. Flints occur in both the Thracian Lowlands and Danubian Plain wine regions of Bulgaria, for instance, and in the Marmara wine region of Turkey's Thrace peninsula.

'Kefraya', to wine enthusiasts, is the name of the winery in Lebanon's Beka'a Valley, born in wartime but surviving to produce what arguably is the country's benchmark wine. But to archaeologists it's the term for the special kind of flints found there. Unusually large Neolithic flint workings extend about two kilometres (a mile) northwest of the Chateau Kefraya winery.

Returning to France, away from the Champagne–Loire swathe of flinty bedrock, the notable occurrences of flint are as loose debris, typically in spreads of river gravel. For example, deposits of flint pebbles come and go along the left bank of the Garonne and Gironde, having been eroded from the Pyrenees and transported northwards. You can see them in parts of Graves, say around Martillac, on some of the gravel rises in Margaux, and in Bégadan in the northernmost Médoc. Some of the *beaux cailloux* (beautiful pebbles) at Saint-Julien's Château Ducru-Beaucaillou are flint.

Finally, a sign that you're in flint country, more obvious than looking for vineyard pebbles, can be the local building material. The odd shapes of flint lumps make them an awkward masonry stone, costly and time consuming to work, and hence they were often only used for churches and other buildings of status. How they were worked changed through time.

In East Anglia, for instance, early buildings were of coarse and unsplit flint alone but at a later time it was combined with other materials, in fancy bands or chequer patterns. Then it became fashionable to trim the flints to square surfaces so they could be framed, so-called flushwork. The various patterns have names like 'concentric', 'bird's beak' and 'snail creep'. It's a whole different world of flint!

Flint in vineyards

You can bet that most growers who have flinty vineyards are going to say so in their marketing, in just the same way as other producers trumpet their particular geology. It seems to me, however, that having flinty soils is something of a mixed blessing. First, the advantages. It sounds good, and lends itself to promotional prose. And in particular, the popular idea that flinty soils imbue wine with a special flintiness is an effective and frequently used selling point. Maybe they do, but it would have to be in some complex, indirect way. Vines are demonstrably incapable of absorbing and using a solid, inert compound such as silica, which is tasteless anyway. This isn't to dispute that flinty wines are produced from flinty soils; the issue is to what extent it's the soil that's responsible. Clearly, numerous producers with flint are convinced it's the key, but on the other hand plenty of flinty wines are produced on soils without flint – Chablis is an obvious example – and vice versa.

The other advantage that's frequently claimed for flint pebbles is their absorbing heat during the day and re-radiating the warmth at night. But again, growers with other kinds of rocks make the same claim, and anyhow it's of doubtful benefit (*see* Chapter 1). The scientific data show that this thermal behaviour varies little between differing rock types – all rocks do it (provided the ground is bare) – and that it's not a very great effect in practice. It's probably only significant in some cool-climate areas where the grapes are trained close to the ground. In any event, there is a school of thought that finer grapes are produced where nighttime temperatures are markedly cooler than during the day.

Flint pebbles provide good drainage and, being inert, they restrict nutrient availability, giving meagre soils that tend to deliver finer grapes. At the same time, in overly dry areas the pebbles can promote the localized formation of dew, which leads to increased microbial activity and more organic matter beneath the stones. All this promotes a good soil structure, though it's the existence of the pebbles that does the job, not that they're made of flint.

The disadvantages arise from the practicalities of working the vineyard. I'll discuss shortly how flint can have sharp edges, and in the vineyard this can lead to punctures. Flint pebbles are tough, and can damage farm machinery: there are plenty of stories of broken rippers, twisted mower blades and dented cultivators, all of which cost time and money. Henry Laithwaite enthuses about his Buckinghamshire vineyard's flinty soils but bemoans that they 'make it a nightmare to work and cultivate – as two broken harrows testify'. The problem is reflected in the vineyard's name: Harrow and Hope.

Moreover, when the steel of an appliance strikes a flint pebble there may well be a spark, and in dry conditions a fire could be started. In fact, just last year, the Harrow and Hope vineyard suffered fire damage from sparks that took hold in dry grass. At Domaine de la Chezatte in Sancerre, water buckets are scattered throughout the vineyard ground in case machinery sparks a fire.

So, flint has a certain allure but practical snags: pros and cons. Rathfinny vineyard in Sussex, England, tackled the dilemma by removing flint chunks from the vineyard – more than a hundred tons of them – and then using them as highlights in the construction of new winery buildings. So now their B&B guests, rather than ponder in the vineyard on the value or otherwise of stones, can delight in the charisma of flint in the walls of their accommodation.

Flint in wine

Now we come to the heart of the matter: flint in your wine. Some of the most frequently used words in today's wine descriptions are 'flint', 'flinty', 'gunflint' and the like. And no doubt all wine lovers will know precisely what the terms mean to them. However, trying to explain the meaning to others begins to reveal some formlessness to the words. Describing a wine as showing a 'whiff of gunflint', for example, seems precise and unambiguous, but as already indicated, on examination it can't possibly display such aromas.

An immediate difficulty is the inertness of silica, such that geological flint is without taste or smell. That's why diatomite – made of silica – is widely used in wine filters: it doesn't taint. It's why silica is used for wine glasses and bottles: in trade-speak the clear bottles used for white Bordeaux wine and some spirits are known as 'flint glass'. So when we use flint to communicate a taste perception, it can't be literal: there has to be something more metaphorical going on.

Flint is most often used to describe acid, perhaps cool-climate, white wines, though it isn't restricted to them. For example, the 'Cayuse Grenache has an abundance of concentrated dark fruit and an unmistakeable flinty streak'; 'Chave Hermitage has a flinty, velvet texture, and almost weightless concentration'; red 'Priorat is known for bold and always flinty wines'.

There is a plethora of definitions out there on what flint means in describing wines, and debate on whether it is a taste, aroma, texture, mouthfeel or some combination. I suggest that three aspects of flint, the geological substance, are at the root of things, and I will look at each in turn.

Flint has the edge

What do the following have in common? The Les Chailloux (Sancerre) is 'wonderfully chiselled and flinty'; 2016 Le Craie, Vouvray, shows 'a flint edge that seems angular'; the 2015 Chablis Vaillons 'has a touch of flint-lending edginess'. It seems to me that they're all alluding to the physicality of geological flint, and its well-known propensity to break with angular edges.

We've all seen pictures of prehistoric flint tools, and perhaps admired their exquisite craftmanship. But here's an extraordinary thing. For ages, flint was

OPPOSITE: The dramatic (and unusual) sunset colours of this piece of flint result from the presence of impurities such as iron; from Nethers Flint Quarries, Flint Ridge, Ohio, USA.

absolutely central to the livelihoods of our ancestors. They – or seeing as they're our forebears and we're all the same species, *we* – knew exactly where to get flint and how to laboriously fashion it into the tools that were so essential. We did it for more than 99 percent of our history.

Then suddenly, in Europe at least, we simply forgot all about it. It simply vanished from memory. So when in medieval times people started noticing flint arrowheads on the ground we were at a loss as to what they could be. Most ideas involved their having fallen from the skies, and in those religious times a common belief was that they were debris from celestial battles involving Satan.
It wasn't until travellers returned from abroad having seen flint still in use that we rediscovered the real nature of these objects.

That was far from the end of it, though, as more troubling questions then emerged. In places these human tools were found buried beneath several metres of sediment that had accumulated through time, alongside the bones of mysterious creatures such as mammoths and the European rhinoceros. How could this be? Humankind was a creation separate from nature and had been placed on Earth only recently – few at the time questioned that. The arguments that followed were impassioned and prolonged. Scientists had to be particularly cautious with such sensitive things, especially in a field beset by fraudsters – remember the Piltdown epic?[1] Even by the time of the *Origin of Species* (1859), Darwin felt it safer just to side-step saying how humanity fitted in.

And here's another, I think, intriguing thing. Archaeologists used to think that this ability to craft tools was the defining feature of modern man, *Homo sapiens*, so it could only have been done during the last 200,000 years or so at the most. Now, not only have manufactured tools been found dating back well over *three million years*, and produced by several different species of our hominid predecessors, but they have also been observed being made, albeit crudely, by chimpanzees and bonobos.

All of this comes about because of the particular way in which flint breaks. It's that three-dimensional crystalline meshwork again! A fracture curves as it spreads through a solid unless there are guiding weaknesses, which with most materials is the case. Silica, however, with its efficient internal framework, has no such

1 The bone fragments discovered at Piltdown, Sussex, became widely accepted for many years as representing a previously unknown type of human, so-called 'Piltdown Man'. When the bones were shown to have been fabricated, this 'Piltdown Hoax' became a cause célèbre.

weaknesses. Consequently, all silica materials, including flint, quartz and glass, break in this curving way. Two curves meeting each other produce an edge and three or more give a point.

I'm not suggesting that a flintiness in wine simply indicates sharpness or acidity, but perhaps, subliminally at least, this edginess is part of the context. Incidentally, that figure who is something of a modern messiah to some in the wine world, Rudolf Steiner, pronounced specifically on how silica breaks. Most of his proclamations are mystical and untestable, but during a talk on the Ego and Meditation he stated: 'Take a quartz crystal. If you take a hammer and break it up, the single pieces retain the tendency to be six-sided prisms, capped by six-sided pyramids.' (He went on to assert that from the pieces 'something living and cosmic' emerges.) But manifestly quartz crystals don't break like that; it's easy to demonstrate that they break into curving, shapeless fragments, just as with glass.

Toilet tip-off

Most instances of flint in tasting seem to be referring to its *aroma*. However, because geological flint has no smell this cannot be literal: geological manuals list more than a dozen technical properties of flint (colour, hardness, translucency etc) but none of them mention smell. So more than likely the references are to the distinctive odour of two lumps of flint being rubbed together. Indeed, I read in a popular UK wine magazine that 'if you want to experience what flint smells like, next time you are walking in the South Downs, pick up two pieces of chalk and rub them together'. Wrong material (chalk is a very different thing from flint) but right idea.

Scientists now have a grasp of how the smell of abrading flints comes about, thanks to their selfless quest for a better understanding of the world around us. A Swiss team was studying toilet malodours – yes, that's right: toilet malodours – when it accidentally isolated 'a flint-like odour, similar to the smell associated with a dentist drilling teeth'. They were able to show that it was due to one compound in particular: hydrogen disulphane or HSSH. Natural flint always contains impurities (which is why its colours are so varied), among them carbon and sulphur. And it turns out that from these and the energy input from rubbing the flints together in air, tiny amounts of HSSH gas are generated from the flint surface.

But here's the thing. A panel of wine professionals tasted various Swiss Chasselas wines and judged that two of them were far more flinty than the others

– and those two contained significantly more HSSH than the others. So if we call a wine flinty it may well contain HSSH and, subconsciously or otherwise, we might be recalling the odour of striking flint pebbles against one another. It's a mental association. It makes sense to me, except I'm not sure how many tasters happily describing a flinty note in a wine will actually have banged two pieces of flint together. I suspect that, as with a number of other wine descriptors (*see* Chapter 3), we *imagine* what the smell should be like.

Other studies have related flinty aromas in wine to the presence of another sulphur compound, called benzenemethanethiol. Apparently, we can detect this in astonishingly tiny amounts – as low as 0.3 parts per trillion – and levels of more than 10 parts per trillion have been measured in some Loire Sauvignon Blancs. The rubbing of flints can hardly be happening in a grapevine, of course, and these sulphur-bearing compounds aren't present in grapes. They're generated during the vinification, and especially during the yeast fermentation phase. It's a further illustration that although we may be labelling our perception as flint, there's no direct connection with any flint that happens to be in the vineyard.

No doubt some winemakers seeking to increase flintiness in their wines will be tempted to nurture these sulphur-bearing (reductive) compounds. For instance, they could eliminate stirring of the lees, minimize racking and avoid new oak. Or the winemaker could go online and, for a mere €191, buy a gram of benzenemethanethiol…

The great gunflint saga

A third variation on sensing flint in wine, besides sharpness and the aroma of abrading flint, is that of referring to 'gunflint'. So we're now in the realms of flint and fire, sparks and smoke. And famous is the association of flint with sparks and making fire, though there are an awful lot of hazy (or should that be smoky?) misconceptions about it. Let's be clear: bashing two pieces of flint together will not make a spark; flint cannot spark. Flint doesn't 'contain fire' as one poet had it. It all comes back once again to that inertness of silica: flint cannot vaporize (and so, as we have seen, lacks odour) and therefore it cannot ignite. So where do those legendary sparks come from, if not from flint? The story is fascinating.

Early man exploited fire. For a very long time this involved manipulating natural fires to our advantage, until at some relatively recent time we learned how

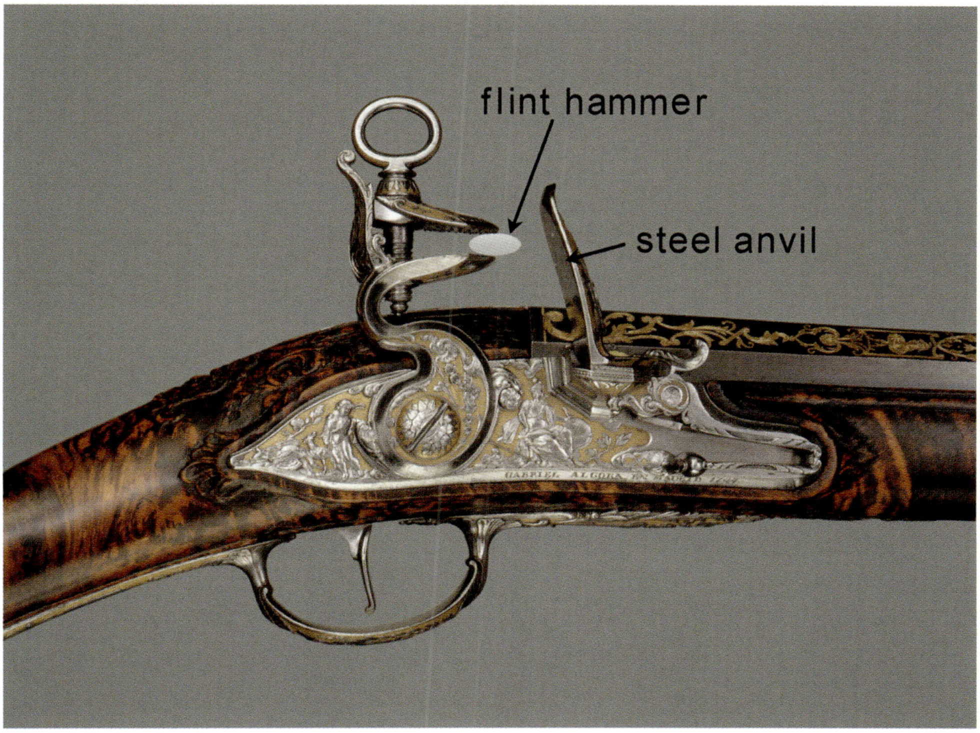

to *create* fire. At first we used the frictional heat from rubbing woods together, and then by striking the mineral pyrite (iron sulphide) on a harder material such as flint, which sometimes makes a spark.

Everything changed when we learned how to manufacture metals. Discovering bronze signalled the end of the Stone Age and our dependence on flint tools, and then making iron (which invariably contained impurities such as carbon, so in practice it was steel) finally gave us an effective way of making sparks and hence creating fire. We soon learned that striking steel on something harder – and flint was familiar and ideal – almost invariably gave sparks. Then with some dry tinder we could create fire. *But it is the steel that sparks, not the flint.* The smell is burning steel.

It is true that striking pieces of flint together in a darkened room may give a visible pulse of light. But this is the curious phenomenon of triboluminescence,

ABOVE: Detail of a flintlock rifle. A piece of flint is held between jaws and, on firing, pivots clockwise to meet the hinged steel anvil. The cover below opens, allowing the sparks to fall into the powder chamber.

in which the energy of breaking chemical bonds is converted into light. That sounds all very technical, but the effect is easily seen with familiar materials – try yanking apart pieces of adhesive tape or crushing a sugar cube. The point is that this doesn't produce heat, nothing burns – there is no smell.

Certain substances in the presence of oxygen are capable of spontaneously bursting into flames. That is, they are pyrophoric. Sodium, potassium and calcium, for example, are extremely pyrophoric. It's why, if you can recall these elements in your school chemistry lab, they would have been kept under a liquid, to exclude oxygen and avoid spontaneous combustion. The phosphorus used in match heads needs the addition of a little heat, such as from the friction of striking a match. The mineral pyrite and a few metals – iron is one – can be struck such that the surface area of any tiny particle that breaks off becomes exposed to sufficient oxygen for it to auto-ignite, thus making a spark. And that tiny speck of burning iron makes the distinct burning smell.

It's easy to get steel to do this by striking it on some hard, tough material – like flint. And if you've ever done it, it can be useful to remember that smell when trying to describe a wine. But clearly in wine tasting it must again be an allusion, a mental association. Vines and wines don't involve bashing and combustion.

Historically, things entered an ominous new era when gunpowder appeared in Europe. When ignited by a spark, this amazing powder burned so rapidly that if it was confined in a metal chamber so much pressure was spontaneously generated that a projectile could be ejected at very high speed. Oh dear, we now had guns. All sorts of ingenious devices appeared for striking steel on a flint and getting the spark to drop into the gun's powder chamber. Then in around 1600 a spring-loaded mechanism internal to the gun was invented – the flintlock. It greatly streamlined things, and became commonplace.

And as a result, flint mining suddenly underwent a staggering new lease of life. The great flint centre of Le Grand-Pressigny was resurrected, together with the extension of the Grimes Graves deposits at Brandon. These two places, in France and England respectively, may not be well known but they quietly underpinned a great deal of history. Their flints armed the pistols of the English Civil War and the French Revolution, the muskets of Waterloo and Crimea. Patriots in the American War of Independence relied on French ammunition: in a way, the Revolutionary War was fought between flints from Pressigny and flints from Brandon.

So I suppose the aroma of gunflint would have been familiar to the military folk of those times, together with hunters, felons, buccaneers and the like.

However, I can't help wondering, especially seeing as modern firearms (together with barbecue, stove-top and cigarette lighters) use synthetic rather than geological flint, how many of today's wine commentators who blithely refer to a 'gunflint' aroma have actually experienced it. But in any case, there's a bigger problem with this much-loved gunflint analogy. The overriding smell of a firing gun – antique or modern – is the reek of the burning powder. The odour of the spark within the confines of the gun chamber is lost in the smell of the exploding charge. It's one thing to strike a piece of steel on a lump of flint and smell the spark, but the odour of gunfire is different.

It seems to me that gunflint as a tasting term isn't a direct comparison with firing guns, but is a nice-sounding variation on the 'smoky', 'struck match' theme. Thus tasting notes speak of 'gunpowder-like smoky hints'; 'as flinty as they come, with the smell of fireworks'; 'intense aromas of gunflint and black rifle powder'. And we may also be subconsciously involving a fair dose of imagination on what a gunflint *ought* to smell like. In turn, and extending this imprecision, this smokiness blurs into the rubbed flint aroma of the previous section.

Whatever we mean by these various labels, there's also the much-mentioned idea that they come from the soil. Things like 'the flint in the soil gives the wine an aroma of gunflint'. It sounds straightforward enough at first, but as I have indicated there are scientific difficulties with the notion. And how it would work is never said. Just how do obdurate, inert flint stones in the ground give 'flintiness' to the finished wine?

I began by saying that everyone has heard of flint. But I wonder how many appreciate how central the role of flint has been in the human journey. As we have glimpsed, flint has coloured literature, influenced architecture, and underpinned warfare. For three million years and more, human technology was dominated by this stuff, our livelihoods depended on it. You could even argue that flint is the single most important substance in all of human history. Oh, and today it's one of the most popular words for describing wine.

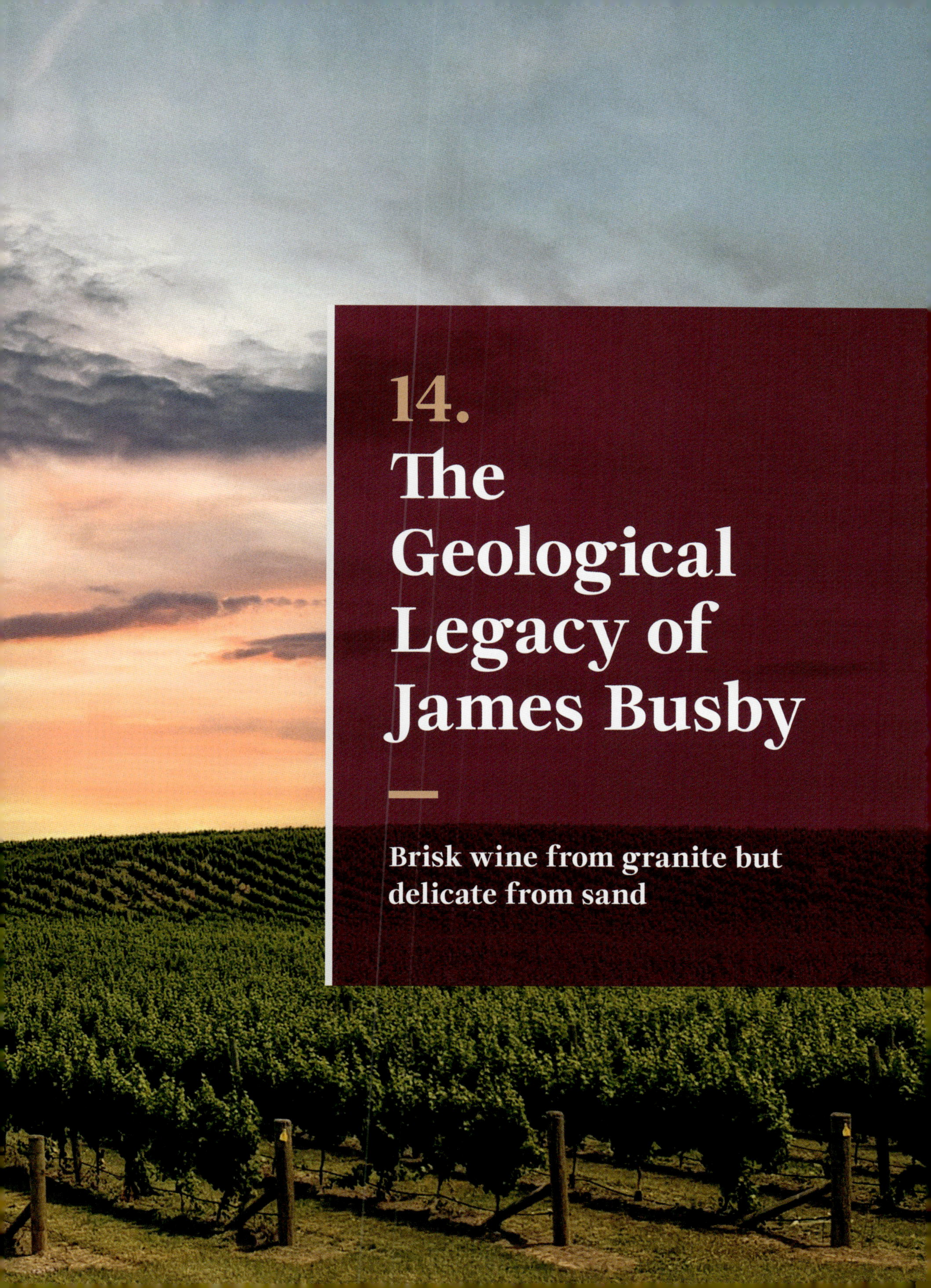

14.
The Geological Legacy of James Busby

Brisk wine from granite but delicate from sand

The Geological Legacy of James Busby

Anerley, a suburb of southeast London, is a modest, unassuming and unremarkable place. However, in the district cemetery is the grave of James Busby, an individual who in life had shown anything but those qualities and who achieved rare distinction. In Australia, for example, he had been pivotal in establishing a viable wine industry, and he is often said to be its 'Father'. In New Zealand he is seen as one of the founders of the country itself, involved with shaping the template that underpins its society today. More widely, he is best known for some of the most influential early writings in English on how to grow vines and make wine.

Busby happened to be staying in Anerley when he died, far from his adopted Antipodean lands, although in some ways the place is fitting. It took its name from the first house to be built in the area, by an expatriate Scot. Anerly is a Scots dialect word meaning 'alone', 'solitary' or 'lonely': Busby was not only himself a Scot by birth but he led a troubled life in which he never really settled or made friends. Nevertheless, he had far-reaching influence: according to the simple upright gravestone, Busby 'drafted the Treaty of Waitangi' and 'introduced the vine into Australia', ending with 'And their Work Shall Follow them'.

Those first two statements are contestable but the closing phrase is certainly true, at least of his vineyard manuals. They inspired the infant wine industry in the English-speaking world and beyond, and even today his writings are cited, especially his remarks on vineyard soils. I know of a dozen or so mentions in the last few years, usually in support of the belief that geology can be tasted in the finished wine; a 2018 book on wine terroir devotes almost a whole page to Busby's words.

I, myself, find it curious that things written almost two centuries ago are being used in modern non-historical works; in fact, this is what piqued my own interest in Busby. Much continues to be written about different aspects of this complex man but I'm not aware of any evaluation of what he wrote about vineyard geology and soils. That is the focus of this chapter.

Formative years

Busby was born in Scotland, within sight of Edinburgh Castle. His father was a jobbing ground engineer, born in northeast England but now assisting with developments in and around Edinburgh. (One of them was at the city's Royal Botanic Gardens: little would he know that today, around two centuries later, in the equivalent Royal Botanic Garden in Sydney there would be a café called 'Busby's Bar'!)

Busby drifted somewhat as a youth, doing a bit of farming in Ireland, dropping out of university after one year, even working as a haberdasher. Through helping his father he became interested in horticulture and then in what today we would call agricultural economics. Which crops were most profitable to grow, in a particular climate?

Aged just 17 he had the confidence to pursue the question by visiting France, renting a cottage in Cadillac, across the Garonne from Sauternes. There he became convinced that wine grapes are, given a suitable climate, one of

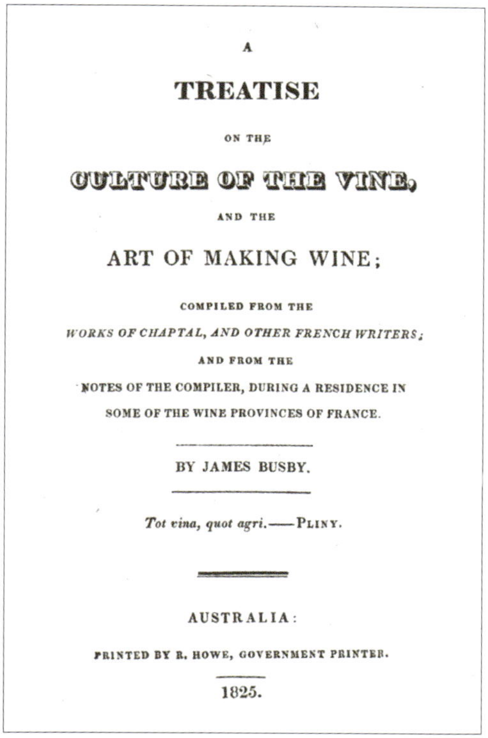

A

TREATISE

ON THE

CULTURE OF THE VINE,

AND THE

ART OF MAKING WINE;

COMPILED FROM THE

WORKS OF CHAPTAL, AND OTHER FRENCH WRITERS;

AND FROM THE

NOTES OF THE COMPILER, DURING A RESIDENCE IN SOME OF THE WINE PROVINCES OF FRANCE.

BY JAMES BUSBY.

Tot vina, quot agri.—PLINY.

AUSTRALIA:

PRINTED BY R. HOWE, GOVERNMENT PRINTER.

1825.

ABOVE: Portrait of James Busby, and the frontispiece of his famous treatise.

the most valuable cash crops. And from observing some of the vineyards and devouring the current French viticultural writings – and especially the magisterial publications of the great Jean-Antoine Chaptal – Busby began to form his own views on viticulture.

In 1823 the Busby family emigrated to Australia, Busby senior having been promised an appointment to deal with mineral and water matters in the growing colony. During the Busbys' long sea journey (accompanied by two milking cows, sheep, pigs, chickens and, intriguingly, cases of wine) young Busby began to write his first major work: *A Treatise on the Culture of the Vine and the Art of Making Wine*. Along with sundry other things he had written during the journey it was published in Sydney in 1825, just a year after his arrival. Busby also brought with him vine cuttings that he had collected en route, during a stopover at the Cape of Good Hope.

But Busby's arrival in the new colony was hardly messianic. Vines had been imported before and there had been earlier writings in English on the possibilities of Australian viticulture; moreover, some settlers disliked the pompous tone of the new arrival's pronouncements. It probably didn't help that the expensive and verbose Treatise was, in Busby's own words, meant for 'the higher classes'. Then as now, Australians didn't like ostentation.

One Sydney critic wrote that the instructions were 'like putting a sum in compound interest before a child who had never learnt the multiplication table'. Moreover, the vines he had brought soon perished. And it seems that Busby's prickly and arrogant nature – traits that were to dog his whole life – soon began to show.

No sooner had Busby managed to contrive a position teaching agriculture at the Male Orphans School near Sydney than he was making enemies and was sacked. He proceeded to challenge this perceived injustice; the start of a preoccupation which was to last the rest of his career. Now unemployed, Busby was driven to try his hand at trading cutlery and the 1828 Sydney census reports his employment as 'tax collector'!

To the ambitious young Busby this was ignominy. He resolved to return to Britain and appeal his job dismissal from the school in person at the Colonial Office, and to scheme for an official appointment of proper status. This decision had two important consequences for the story here. First, the land that his family had acquired in Australia was developed as a vineyard in his absence, and second, it allowed him to pay a second visit to France.

Kirkton, the 'Busby vineyard'

Because the Busby family were free settlers (as opposed to convicts) with more than £500 to their name they were entitled to a grant of land. The allotted Busby acreage was in the Hunter Valley, north of Sydney, today a playground of weekending city dwellers enjoying the resorts, festivals, wineries producing renowned Semillon wines and so on, but in the 1820s it was largely dense brush forest with only one overland road. Consequently, land close to the river was much the most desirable because of the access by water, and for most settlers the flat, alluvial soils offered fertility and relative ease of working for agriculture.

But Busby junior was interested in planting vines, so what would he have thought about the site? His Treatise emphasized the overriding importance for vine growth of factors to do with climate – and certainly the Hunter Valley was more suitable than Sydney – and that the main thing about the soil is the need for it to be poorly fertile, loose and free draining. So Busby must have realized that the alluvial soils close to the Hunter River were less than ideal. (It is notable that of the half dozen or so vineyards that were planted there in the first half of the 19th century, only one now remains: George Wyndham's plantings at Dalwood. All today's Hunter wineries are elsewhere in the valley, on various soils such as limestone and basalt.)

Nevertheless, the Busbys directed their allotted convicts to build a residence on the land. Father John was busy in Sydney so when James decided to go back to Britain temporarily, his brother-in-law, William Kelman, took over management of the land. And so it was that Kelman, rather than Busby, established a vineyard there, and it was he who gave the property a name. Before Kelman had emigrated from Scotland – on the same ship as the Busbys – the young lad had played on the estate for which his father was the bailiff, a place called Kirkton on the outskirts of Fraserburgh in the northeast of the country. This was to be the name. Sadly, rather like Busby himself, this new Kirkton was to have a troubled history.

Busby on soils

What does Busby's Treatise say about vineyard soils? Noticeably, he says little about geology in the sense of the underlying bedrock. However, he explains in some detail how pebbly, calcareous, volcanic and granite soils all usually have the appropriate physical properties for viticulture. Clay, however, is to be avoided.

Clay soils – Busby actually uses the technical term argillaceous – tend to be cold, he explains, with their coherence inhibiting warmth from reaching the vine roots, just as with rainfall. Similarly, their firmness prevents 'the dissemination of the minute fibres of the roots'.

Some of his remarks are awry – the best Bordeaux vineyards are not on soils with decomposed granite; volcanic soils are not impregnated with fire – but most of his statements are carefully justified. His conclusion summarizes his position: 'The intrinsic nature of the soil is of less importance than that it should be porous, free and light.' All this is closely argued and wholly in line with other writings of the time, indeed with today's scientific views.

But then Busby unleashes a single sentence completely out of the blue and at odds with all that has gone before. In its own single paragraph, Busby declares *ex cathedra*: 'The sandy soil will, in general, produce a delicate wine, the calcareous soil a spirituous wine, the decomposed granite a brisk wine.'

He offers no expansion, no explanation of how any of this might come about. Presumably the ideas are based on anecdote, perhaps on things he heard while in France; the baldness of the statement certainly contrasts with the reasoned approach of the surrounding text. Take, for example, calcareous soils (ie, limey, containing significant calcium carbonate). In an earlier passage Busby had explained how such soils are beneficial for vine growth in that they promote free water circulation, allowing 'numerous tender ramifications' of the root system etc, but he made no mention there of any effect on wine itself.

But now he asserts that they yield a 'spirituous wine'. By this Busby clearly means high in alcohol, because in his section on climate he remarks that warmth promotes 'the production of saccharine matter and generally strong spiritous wines, sugar being necessary to the formation of alcohol'. For his discussion of calcareous soils he had used the example of the Champagne area: he must have known those soils didn't lead to high alcohol. In any case, how might calcareous soils promote a higher alcohol content? Busby just doesn't say.

It is this single sentence that is commonly the basis of Busby's works being cited nowadays, usually to bolster claims about the importance of soil for wine taste. For instance, in a 1977 conference paper the Italian viticulturalist Mario

OPPOSITE: The site of the Kirkton vineyard near Branxton, Hunter Valley, New South Wales, Australia, as it looked in 1914 (*top*) and today (*bottom*). Although often referred to as 'Busby's vineyard', the extent to which James Busby was involved is doubtful.

Fregoni stated ways in which soils influence wine flavour and they are taken almost verbatim from Busby: sandy soils give light, aromatic wines, calcareous soils give high alcohol etc.

In turn, Fregoni's list is much cited by subsequent authors. Yet to my knowledge none of these later authors offers, just as with Busby, anything about the rationale of the claims or their validity. And we should remember that – seemingly another mark of Busby's self-confidence (conceit?) – the lengthy Treatise was based solely on a single visit to parts of Bordeaux and his reading of the current French works.

Before Busby left Australia for London he found time to rewrite his Treatise with a more approachable and practical tone. The change is reflected in the new title: *A Manual of Plain Directions for Planting and Cultivating Vineyards, and Making Wine in New South Wales* (1830), and this version was aimed fair and square at 'the class of smaller settlers'. In it, Busby again asserts that the main attribute of vineyard soils is their being 'loose and open' and mentions that usually calcareous, volcanic and granite soils are very appropriate but that stiff clays or rich alluvium loam are unsuitable. However, there are two differences from the earlier version regarding soil.

First, he mentions that 'many people in the wine countries' believe that the nature of the deep subsoil has a role, both in vine growth and in the wine itself. In the Treatise Busby had barely mentioned what might lie below the topsoil. In the new Manual he does, but then goes on to reject the idea. 'While the truth of this can scarcely be considered as established, I question whether, were it ever so well ascertained, we could derive much advantage from the knowledge of it.' In other words, despite introducing it, he saw little relevance in the bedrock geology.

And second, that much-quoted cryptic single sentence about wine traits varying with certain soil? In this new version, it's absent.

A return visit to France

During Busby's return to Britain he took the opportunity for a four-month tour observing agriculture in France and Spain. Among numerous other things, he visited selected vineyards and carefully amassed vine cuttings. Eventually, Busby was to ship to Sydney's infant Botanic Garden a large collection of carefully catalogued vines, where they were propagated for Australian viticulturalists to use. Sadly, these cuttings again did not fare well in the Garden – in fact, today 'Busby's Bar' is the only visible reminder of his connection there.

However, before their demise duplicates of the cuttings found their way not only to the vineyard at Kirkton, but to numerous other sites in the Hunter Valley and elsewhere in the state. They also went, for example, to the Clare Valley, Adelaide Hills and McClaren Vale in South Australia; to Geelong, Ballarat and Bendigo in Victoria. They were acquired by Dr Penfold for his 'Chardonnay', and by Samuel Smith at the new South Australia estate he was calling 'Yealumba' (*sic*).

Without doubt this collection of vines was Busby's major contribution to Australian viticulture, though discussions on its value continue to this day. For instance, its focus on French and Spanish cultivars meant that grapes from elsewhere, such as the Italian varieties that are now proving popular, were long neglected in Australia.

The outcome of Busby's sojourn in Europe that is of more relevance here is the diary of his visit, with his updated views on vineyard soils. Published in 1833, it is entitled *Journal of a Tour through Some of the Vineyards of Spain and France*, though in fact it covers a whole range of farming matters, from hedging to garlic, sugar cane to silk worms. Regarding vineyard geology, in two ways his views have radically changed. First, he has become convinced of the supremacy of calcareous soils for producing quality dry wines, and second, he now stresses the defining importance of methodology – how the vines are raised and how the wine is made.

Soil volte-face: it must be calcareous

Busby was first struck by the significance of calcareous soils while in Spain's sherry-producing region. He noted that the sandy soils called *arenas* led to inferior wines while the finest came from the calcareous albariza soils. Later he visited Burgundy and on the basis of several localities around Gevrey and Vougeot declared that the calcareous, higher vineyards of the Côte d'Or yield 'the driest and best wine'. The clay content increases downslope and consequently the wine 'falls off in quality till it becomes the *vin ordinaire* of the country'. In Champagne he was again struck by the desirability of calcareous soils, although he conceded that the clayey soil around Aÿ 'in some degree affects the quality of the wine, but not in a great degree'.

But the place of Busby's seeming epiphany is a surprising one: the famously granitic Hermitage hill in the Northern Rhône. There, he made an unplanned call on a local proprietor who told him that the finest Hermitage wine came from a blend of grapes from the 'ordinary granitic soil' with those from 'a belt of calcareous soil' that crosses the hill. Busby obtained some of this latter soil and after seeing it fizz when he poured vinegar onto it wrote that, indeed, it contained 'a considerable portion of lime'. He then proclaims, once again without expansion, that 'it is probably to this peculiarity that the wine of Hermitage owes its superiority'.

In asserting this supremacy of calcareous soil Busby is completely reversing his earlier conclusions, stated in both the Treatise and the Manual. Moreover, he is rejecting outright the view of Chaptal and the other eminent French writers of the time. They consider, Busby reports, that 'provided the soil is porous, free and light, its component parts are of little consequence'. 'Granitic, schistose, argillaceous, flinty, sandy and calcareous soils' are all 'equally well qualified to produce… wines of the finest quality'. This was exactly Busby's previous position; now he is dismissive.

Writers on other aspects of Busby's life have commented on the man's perpetual arrogance and pig-headedness, and that seems to be illustrated here. He declares that Chaptal and company have been deceived. Jean-Antoine Chaptal, *comte de Chanteloup et comte de l'Empire*, once Napoleon's Minister of the Interior, was France's pre-eminent expert on chemistry and agriculture, including viticulture (he owned a vineyard in the Loire), and author of several standard textbooks. Busby had no training and virtually no hands-on experience of either viticulture or geology.

Nevertheless, apparently because of a chance conversation during his short visit to Hermitage, Busby proclaims: 'These writers have in many instances been misled… the fact is that the wine of the hill of Hermitage owes its superiority only to the circumstance of the granitic soil of a part of that hill being mixed with calcareous matter; and but for this circumstance, I am satisfied that the wine of Hermitage would never have been heard of beyond the neighbourhood where it grows.'

He had probably visited the le Méal vineyard, high on the southern flank of the Hermitage hill, which is veneered by carbonate-bearing loess overlying the granite. It's still a respected site today, though just one of several, most of which are located directly on granite. But Busby was convinced. We are not told how he reconciled his new idea with his experience of the gravels of Bordeaux or what he might have thought if his European itinerary had taken him to places with different geology. But as it is Busby could not be clearer: 'The soils of all vineyards producing dry wines of superior excellence are strongly calcareous.'

Busby says nothing about wine character or flavour: he is talking throughout about the *quality* of the wine, *vin ordinaire* versus fine wine. He explains what in practice this means to the French: ordinary wine is drunk with ordinary meals, in tumblers and usually mixed with water; fine wine is drunk pure, and in wine glasses. And, according to the converted Busby, to produce fine wine – anywhere – you need calcareous soils.

But really, it's all in the winemaking

It's surprising that Busby didn't make more of the effect of soil on wine taste because he was writing at a time when plants were still thought to be *made* of soil. (Photosynthesis wasn't recognized for another half century – *see* Chapter 2.)

This was probably Busby's understanding: he wrote of vine roots being 'in search of juices which are appropriated by the plant'. But nothing about the soil itself affecting taste. In other words, modern works that cite Busby supporting the notion that geology can be tasted in the wine have to derive from that terse and anomalous single sentence back in his 1825 Treatise.

However, despite all this, Busby followed his pronouncement on the supremacy of calcareous soil by then going on to play down the importance of vineyard soil. Busby detailed how true wine excellence is achieved by careful management of the vineyard, optimal harvesting and assiduous attention to vinification and cellaring. This is how 'the vineyards of a few individuals acquired a reputation which has enabled the proprietors to command almost their own prices for their wine'.

Today's marketing commonly promotes the specialness of a wine because of the uniqueness of the soils – about the only factor that can't be replicated – and back in his time Busby felt that this same idea was being used to deter rival producers. But, he says, 'there is more of quackery than of truth in this'. He continues, 'it was evidently the interest of such persons that the excellence of their wines should be imputed to a peculiarity in the soil, rather than to a system of management which others might imitate'. *Plus ça change…*

The mirage of Kirkton, 'Busby's vineyard'

Presumably, Busby's pronouncements on vineyard soils carry added weight today because of the perceived stature of the man as a pioneering wine authority. But in view of Busby's dearth of actual involvement with wine, it's as though with time a veneer of mythology has grown up around him. The thought is illustrated by the celebrated 'Busby's vineyard' at Kirkton, regarded by some as the cradle of Australian wine.

In 1828 William Kelman planted vines at Kirkton, on the higher, less clayey parts of the estate, presumably inspired and guided by his brother-in-law's writings. In 1830 he produced wine and the business took off. In time, Kelman's son James took over, and won awards both in Australia and abroad for his wines. Apparently, although James himself was teetotal, they were 'very inebriating'! He established the much-loved local tradition of inviting folk to celebrate each vintage at Kirkton with a banquet, picnics and 'athletics events with prizes'.

These were the glory years of Kirkton, and in a way an expression of Busby's vision of the viticultural possibilities in Australia. In 1914 the estate was bought by the respected Lindeman family of nearby Cawarra and in 1930, with some of the original plantings still thriving, the Lindemans held a grand celebration marking 100 years of winemaking at Kirkton. A hundred gallons of Kirkton 'Chablis' and 100 gallons of 'Burgundy' were served; they even managed to serve some bottles from the very first vintage. But that was to be the swansong of Kirkton.

The following year the Lindemans decided to pull out of the Hunter Valley and the Kirkton land then passed through a series of owners, none of whom were interested in viticulture. The heritage vines were pulled up and the house demolished. All that was left was the cellar. Its floor still had some of the original bricks once made by convicts from a nearby red clay, but not long ago even these were taken away, for some ignominious use elsewhere. All that is visible today among the rough pasture is a couple of trees and a single flat grave slab, weathering badly but with some Busby and Kelman family names just visible. Even the very name 'Kirkton' has now been usurped by another property.

But in a way, that there's hardly anything to see there today adds a veil of mystique to the place, an invitation to use the imagination. These days (according to the present owner) the site of 'Busby's vineyard' is a place of pilgrimage for wine enthusiasts from all over the world. After all, 'James Busby's name is synonymous with all that is great about Australian wine', one web-page proclaims, and there's a Kirkton Wine Club, 'named in honour of… Kirkton' and 'its intrinsic association with James Busby'. It all enhances the status of Busby, and hence the gravitas attached to his words.

Unfortunately, the written records suggest a different picture. The Kirkton vineyard was, as we have just seen, planted and operated throughout by the

Kelman and Lindeman families. Successful it may have been but so were plenty of other Hunter Valley estates. Nor was it the first. George Wyndham, for example, planted vines just to the east at the same time and Thomas Webber was producing wine less than 20 kilometres (12 miles) from Kirkton in 1828.

What is often referred to as James Busby's land and James Busby's house actually belonged to his father, John Busby, who retired and died there. Hence his and the Kelman names on the still existing grave slab. James Busby was later allotted land fairly near Kirkton but there is no record of his ever having done anything with it, apart from selling it on in 1832.

Conceivably, Busby visited family at Kirkton, though his family relations were often strained. It would have been a time-consuming journey from Sydney for the busy man: surely there would be a mention in his fastidiously detailed writings? And Busby felt Kirkton unsuitable for viticulture. He wrote to a friend (Busby was an inveterate letter writer) that: 'I cannot say that I think it prudent of Kelman to launch into a plantation of 10 acres of vines before he has proved the capabilities of the soil. I think it unlikely that his wine will ever be above the Vin Ordinaire.'

The fact is there is no record of Busby ever having been to Kirkton, let alone working vines there and making wine. (There seems to be no record of his having made wine anywhere.) Nevertheless, Kirkton has been called 'an icon' and 'Australia's most important heritage site'. One visitor remarked that standing at Kirkton paying homage to Busby 'sent shivers down my spine'. Although these perceptions are hard to square with the evidence, Busby himself undoubtedly would love them.

Busby tackles New Zealand

Busby's writings on viticulture were without question genuine attempts to encourage and advise the infant wine industry in New South Wales and elsewhere, and, together with the spread of his propagated vine cuttings, they were highly effective. They were, though, for Busby just part of a wider agenda.

One broader aim was to encourage the drinking of civilizing wine rather than inebriating spirits. The very first page of his Manual spells it out: the book's ultimate purpose is to 'promote the morality of the lower classes'. Another aim was to improve agriculture in general in the colonies and so lessen their dependency on the mother country. Hence his advice in the Journal on producing raisins, almonds, figs etc. In other words, Busby saw himself as a colonial

improver. But even this was part of a more fundamental motivation, one that drove Busby throughout his adult life: the need to acquire status, improved social standing – *to be admired*. Expounding on wine was part of this striving.

The trip to London that enabled Busby to visit France and Spain was made primarily for him to try to persuade the colonial powers-that-be that the fledgling, ungoverned and anarchic settlements in New Zealand needed the presence of a person of distinction and authority. Oh, and he could be just that person (even though he had never been there).

ABOVE: The house on the Waitangi Treaty grounds near Paihia, Bay of Islands, New Zealand, from where Busby acted as the country's official 'British Resident'.

His machinations worked, and Busby got himself appointed as official 'British Resident' in New Zealand. This was only eight years after he had first arrived in Australia from Scotland, so Busby, still only 30 years old, had certainly done well for himself. (Though he did have some concerns about his new position: would his house be suitably imposing and built of brick; would his uniform be a properly embroidered and tasselled tailcoat, with plumed hat?)

A great deal continues to be written about the doings that followed Busby's appointment and what it all means today for New Zealand. I merely note here that Busby did plant a 'grapery' next to the residence he had built at Waitangi, in the North Island's Bay of Islands. (Because of it, even though Busby's weren't the country's first vines, he has been dubbed by some 'the Father of New Zealand wine'.) But Busby seems to have no longer been very interested in wine. Among the vast amount of material he turned out during his nearly 40 years in New Zealand there is no mention of wine, viticulture or vineyard geology.

Busby's house still stands, on the Waitangi Treaty grounds, but it presently has just a few straggly vines growing behind it. (For what it's worth, the soil at Waitangi is, just as at Kirkton, alluvium with patches of the clay that Busby had once derided. But fittingly, it overlies a bedrock of what is effectively New Zealand's national rock, greywacke.)

Evidently Busby's cantankerous nature travelled to New Zealand with him, for soon he was making enemies all over again and getting involved in new legal squabbles. And over time, his always vague official authority progressively drained away and he became heavily in debt as successive businesses failed. Moreover, his eyesight continued to deteriorate. Thus in 1871 Busby travelled to London for cataract surgery. The operation was successful but while convalescing he caught acute bronchitis, which soon worsened. Shortly, far from his adopted lands, James Busby died, according to the official record from 'congestion of the lungs'.

There seems to have been no mourning from the other side of the world, scant mention in the newspapers and no memorials. And that is why this significant and influential individual comes to be buried unceremoniously in a suburban London cemetery. And there he lies still, in his solitary grave, disillusioned, unfulfilled, embittered and lonely. Or, as they say in Scotland, anerly.

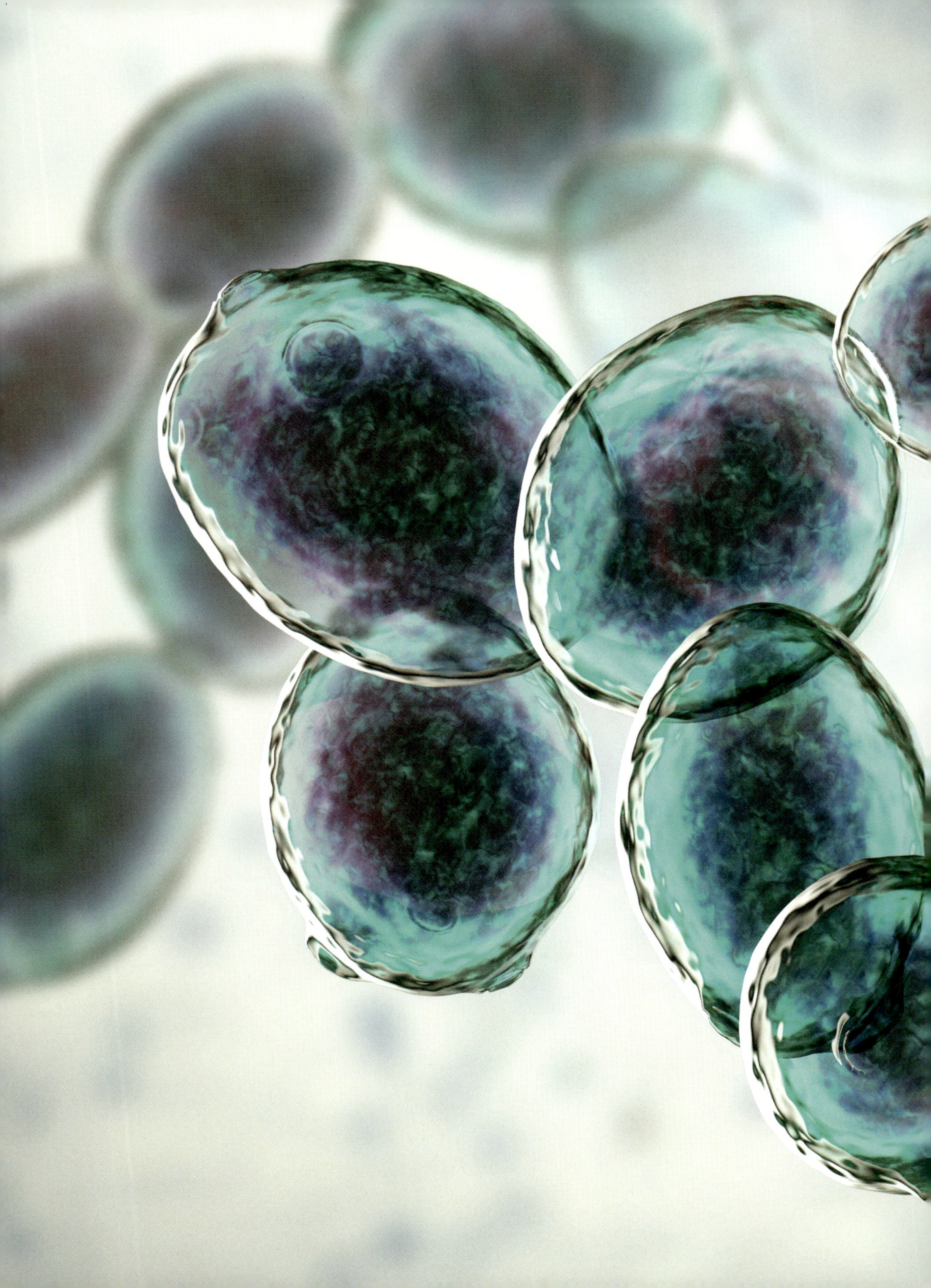

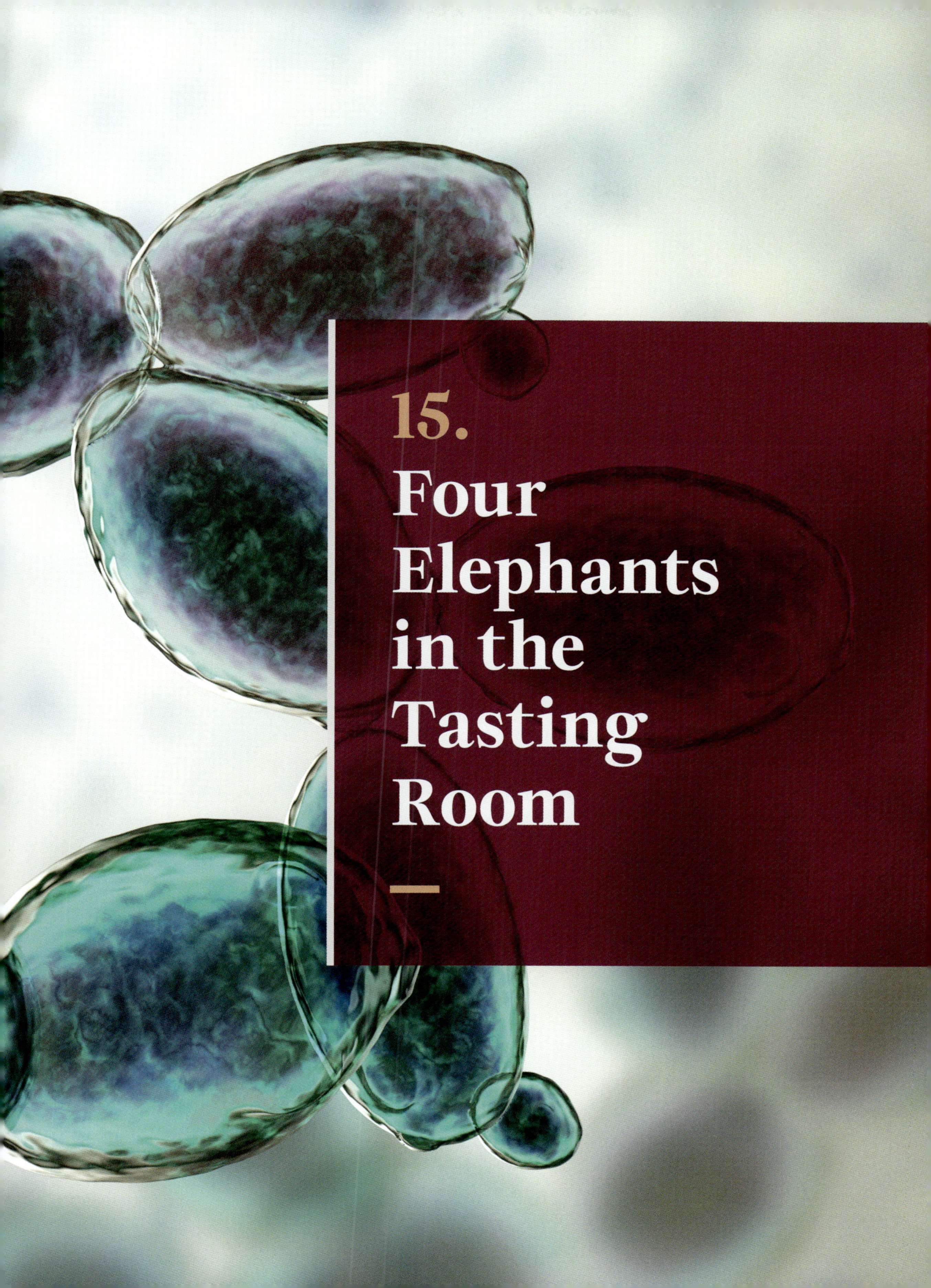

15. Four Elephants in the Tasting Room

Four Elephants in the Tasting Room

1 The goodness of the soil

'The rain quit and the wind got high,' sang Woody Guthrie, 'a black old dust storm filled the sky.' For John Steinbeck: 'The sky was darkened by the mixing dust, and the wind felt over the earth carried it away.' The awful 1930s Dust Bowl on the American prairies is burned into our consciousness, but similar though less storied catastrophes have happened and continue to happen in other places around the world.

Such calamities come about essentially because the soil has lost its 'goodness', that is, its humus, the key to a healthy soil. Humus, decayed organic material, is in practice the main source of the minerals that the grape vine takes up and, as some would have it, reach the wine for us to taste. But when did you last see humus mentioned in a vineyard description or in a wine-tasting note? Wine lovers don't really want to read about rotted insects and decayed plants so, well, it's just ignored. In other words, we have here an 'elephant in the room' (defined in my dictionary as a huge topic that is uncomfortable for people to talk about and so is ignored).

The humus in a soil becomes depleted because of overcropping, which eventually leads to crop failure. If this is accompanied by periods of drought the exposed soil soon becomes crumbly and is easily blown around. Hence dust bowls and the like. A current viticultural example is the overcropping resulting from unchecked demand for Prosecco, with consequent soil erosion problems. One recent study took the volume of wine produced in a year and the amount

PREVIOUS PAGES: Cells of the common wine yeast *Saccharomyces cerevisiae.* Perhaps more important to the character of wine than rocks and soil.

of soil being lost annually and calculated that a single bottle of Prosecco was costing 4.4 kilogrammes (nearly 10 pounds) of lost soil.

Most farmers and gardeners know that you can't keep extracting crops without putting 'goodness' back into the soil. Synthetic fertilizers will provide a temporary nutrient fix but organic matter such as manure or compost will in addition improve the texture and microbial life in the soil. Hence that's probably what most people mean by soil 'goodness'.

In nature, soil nutrients are recycled. In a woodland, say, at the end of the growing season some of the vegetation, including the nutrients it has been using, is discarded. It falls to the ground where it's processed by the little creatures that live there into leaf mould, more generally called compost. Eventually, as the decay is completed, the material becomes the non-living, dark spongy matter we call humus. (Pronounced *hew-mus*. Note the single 'm'. With two, it's the chickpea spread that to humans is much nicer to eat.) And not only does the humus contain the recycled nutrients, they are easily accessible for the next season's growth.

In agriculture, however, as each year's crops are harvested the nutrients they contain are extracted from the system – the cycle is broken. So with successive harvests the nutrient content of the soil declines. Vines have unusually modest needs but even so the grower will need to make sure adequate nutrients are there in the soil, and accessible. This may involve (apart from using artificial fertilizer, which will do nothing for the soil condition) letting vine vegetation, weeds or cover crops decay in place, or spreading modest amounts of compost, pomace or other organic matter onto the soil.

Slowly weathering

The chemical elements that can act as nutrients are locked inside rocks and the geological minerals that make the soil framework and so are inaccessible to plant roots. It's weathering that progressively liberates them. But this involves a whole series of intricate processes that take place slowly – very slowly. We've all seen old gravestones, perhaps hundreds of years old, weathered such that the epitaph is becoming hard to read. In fact, weathering rates are typically measured in thousands of years: hence the processes are simply unable to produce each year a fresh suite of nutrients. And that's where humus comes in.

Because of the recycling explained earlier, humus is rich in nutrients, and in fact it's the only source for plants of vital nitrogen. Moreover, humus is able to

release the nutrients (through what is technically called cation exchange) much more readily than any geological mineral. So although weathered rocks are the fundamental, primordial source of nutrients, in practice, season after season, much of the nutrient provision comes not from the vineyard geology but from the soil humus.

Unromantic reality

Wine is almost wholly composed of carbon, hydrogen and oxygen; the inorganic content – the 'minerals' in a wine – is typically no more than 0.4 percent and most of that is potassium. Despite this tiny proportion, those minerals get much mention in popular wine writing, often with at least an implication that their presence is a result of vineyard geology: 'the granite soil offers minerals which impact the mineral flavours of the wines'; 'the vine transmits the minerals all the way from the stones around its roots to the final wine'; 'a very mineralized taste coming from the special mineral richness of the volcanic soil', and the like.

ABOVE: These alluvial cobbles of basalt make a very barren-looking soil but around the vine roots there must be sufficient humus. Cayuse Vineyards, Walla Walla, Washington, USA.

But as I discuss in Chapter 3, of that tiny amount of inorganic material in the wine a substantial proportion will have come from sources other than the ground. Nevertheless, a proportion may have originated in the soil and survived through to the finished wine: my argument here is that it almost certainly originated in the humus.

Flint, slate, graphite and the rest might be charismatic and fun to read about in wine magazines, but bacteria, decayed vegetation, rotted creatures, and so on? They're never mentioned. In other words, the real source of most of the soil-derived nutrients, the minerals in a wine, is never acknowledged in tasting notes. Humus is, as I've suggested, something of an elephant in the room.

2 The root of the matter

Franz Richter, Georges Couderc, Federico Paulsen, Antonino Ruggeri... Enthusiasts will be familiar with the names of some great contributors to the wine world past and present, but I can imagine puzzled frowns and raised eyebrows at these ones. I never see them mentioned in the popular wine literature. Yet it is scarcely an exaggeration to say that if it were not for the work of these individuals then wine as we know it might not even exist today.

The vast majority of the world's grapevines are, of course, grafted: the top half being the part we see in the vineyard, with a familiar name like Chardonnay, Pinot Noir, Cabernet Sauvignon, etc, and the bottom half being very largely hidden down in the soil: the rootstock. And most rootstock names come from the people that developed them, thus, for example, Richter 110 (110R), 3309 Couderc, 1103 Paulsen (1103P) and 140 Ruggeri (140Ru). When did you last see these names in a wine magazine?

Not only do rootstocks affect how vines grow and grapes ripen but also they can affect the flavour of the finished wine. Yet they are rarely written about in the popular press: so here, I suggest, we have a second 'elephant in the room'.

A growing question

It's all very different on the professional side of things. In fact, choice of rootstock has been called 'the biggest decision in a vineyard', in that while most other things can be changed or ameliorated, the choice of rootstock is fixed and long-lasting.

So your soils are particularly high in magnesium? Perhaps Malègue 44-53 is the one for you, with its low mineral uptake. Poorly drained soils? Then Monsieur Couderc's 1616 might be best, or his 3309. Limestone soils? 101-14 Mgt (Millardet et de Grasset) is much used for the limey soils of southern Europe.

All technical stuff, pivotal for the viticulturalist, but my focus here is on what these rootstocks might mean for the wine in our glass. It's well established that there are indirect effects on wine, for example a rootstock can promote vigorous growth that leads to poor-quality grapes that are shaded from the ripening sunshine, but what are direct effects?

Conventional wisdom says there aren't any; maybe a view lingering from the time of the phylloxera crisis when rootstocks were anxiously promoted on the basis that they had no taste effects (unlike hybrids), and a nostalgic feeling that wines from pre-phylloxera, ungrafted vines remain unsurpassed. Perhaps this goes some way, as well as the names hardly tripping off the tongue, to explain why rootstocks are so rarely mentioned in the popular press.

But for some years now, scientific evidence has been accumulating about their direct effects on flavour. Here are a few examples. In one experiment in South Australia's Barossa Valley, wines from vines grafted onto 110R or Schwarzmann rootstocks, relative to those grown on their own roots, had a more intense body colour and darker berry aromas as well as more intense and coarser tannins. They were marked higher in blind judgements on quality.

At Cehegín in Murcia, Spain, various rootstocks of Monastrell (Mourvèdre) showed 'a very significant effect on wine composition'. For example, wines from 140Ru were relatively higher in tartaric/malic acid and anthocyanin/tannin ratios but lower in polyphenols and higher alcohols.[1] They were rated higher by tasters.

Shiraz vines grown at Mildura, Victoria, gave wines richer in 'green' flavours such as bell pepper and green beans (methoxypyrazines) where grafted onto Ramsey and Dog Ridge rootstocks, rather than Merbein 6262. Paulsen 1103 rootstocks, however, consistently produced wines with 'dark fruit' characters.

Those descriptions of wine character – dark fruit, green flavours, etc – are the kinds of thing we see all the time in wine reports, but the names and acronyms of rootstocks, well, although they are highly relevant they seem quite alien. But there's a further reason why rootstocks should be on the radar of a wine enthusiast.

1 Higher alcohols have higher molecular weights and boiling points than normal ethanol. They form during the fermentation of sugars and amino acids in grape juice and can affect wine aroma.

The dark side

At San Clemente, in the Maule Region of Chile, experiments on Merlot vines gave results exactly in line with those mentioned above. For instance, vines with 140Ru rootstocks gave wines with enhanced fruity esters whereas SO4 produced a high content of total fatty acids and a low total of higher alcohols. Those vines that were left on their own roots gave the lowest content of esters but the

ABOVE: A Shiraz vine, more than 150 years old, on its own rootstock at Seppeltsfield winery, Barossa Valley, Australia. Seppeltsfield was established in 1851 – the estate was first planted to vines when it was found that the land was unsuited to the more profitable crop of tobacco.

highest content of ketones and higher alcohols. (Enhanced higher alcohols, the experimenters reported, 'confer spirit-like and animal attributes and suppress the aroma of red fruit and woody notes in fruity and oak-aged wines, respectively'.)

But wait! This is famously phylloxera-free Chile so why even bother with grafted rootstocks? Indeed, phylloxera has not been detected there (yet?)... The thing is that *Vitis vinifera* is susceptible to a whole range of other pests and diseases that can seriously compromise performance. In Chile, particular problems are with *Margarodes* (which the French call *cochenille du Chili*), the South American fruit tree weevil and a wide range of nematode species. Traditionally in Chile, these were controlled by soil fumigation and spraying with agrochemicals, procedures that are no longer environmentally acceptable or, in some cases, legal. So, perhaps as with phylloxera, grafting onto resistant rootstocks is the best defence – but even then, there are problems.

Some common rootstocks

"name"	vigour		resistance		soils				
	low	high	nematodes	drought	shallow clay	deep sand	saline	acid	alkali
riparia x rupestris									
3309 Couderc	✓		X			✓		✓	X
101-14 Millardet			✓	X			X	✓	X
berlandieri x riparia									
161-49 Couderc	✓		X	X	✓				✓
S Oppenheim 4				✓					✓
berlandieri x rupestris									
110 Richter			X	✓	✓	✓		✓	
1103 Paulsen		✓	X	✓	✓	✓	✓		✓
vinifera x berlandieri etc.									
41 B			X		X		X	X	✓
Fercal		✓		X				✓	✓

blank box indicates no marked property

ABOVE: A selection of some common rootstocks and their properties.

The gathering storm

The European phylloxera crisis subsided towards the end of the 19th century as grafted vines finally became established. (I recall in Chapter 11 the heroic efforts of desperately trying to match resistant rootstocks with the soils of southern Europe.) By then, about a score of the newly developed rootstocks had become the most widely practicable – and they are largely the same rootstocks that the world's growers use today. But nature has moved on.

Environmental conditions shift, especially in these days of climate change. A rootstock that used to cope with some aridity, for example, might now be inadequate for today's increasingly intense droughts and soil salinities. Then there's the pests. The predators and pathogens in soils are constantly changing – can the traditional rootstocks repel their new forms? And as for phylloxera itself (putting aside its quite bizarre sex-life), the louse has complex and variable lifestyles that equip it well to adapt to new conditions.

Yet around 99 percent of all vine rootstocks currently used commercially are derived from those same few historical varieties. Consequently, it is a very limited gene pool, which, coupled with the intense monoculture of most vineyards, makes vine roots very vulnerable to their evolving adversaries. In other words, to caricature things just slightly, vines are facing an array of constantly adapting enemies while relying on defences from more than a century ago.

Some vine scientists think an answer may lie in the multifarious wild vine species that extend right across Asia. Others feel that trying to wring further tweaks from rootstock interbreeding should be abandoned altogether in favour of modern methodologies. The obvious and potentially most powerful one is genetic modification (GM). Of course, the very idea strikes horror into many across the wine industry. But then, once upon a time, to many vignerons, so did the notion of adulterating precious French vines with American roots…

3 Choices in ferment

Each autumn I enthusiastically pick the grapes from the couple of dozen vines that I have, and carefully press them. I am then faced with a decision. What kind of yeast should I use to ferment the juice? Even at the level of this obviously very small-scale and strictly amateur hobby I have a considerable choice of yeasts to buy: more than 50, I reckon. The decision is important, because the one I choose will help determine the finished wine: its style, flavour profile, alcohol content, etc. So imagine the weight of the decision facing the commercial producer, with a vast range of possible yeasts to choose from and so much at stake.

I could just leave my grape juice to its own devices and rely on so-called 'wild' or ambient yeasts. But in a place like Wales with so few vines around, that would be a haphazard business. There will, of course, be plenty of yeasts in the air but there's unlikely to be much *Saccharomyces cerevisiae*, the classic wine yeast. I could use some of my beer yeast or purloin some of my wife's bread yeast, but goodness knows what kind of wine they would give. So, over to the suppliers' catalogues.

I first have to decide whether to buy the yeast in dried form or as a liquid. Professional winemakers have all sorts of further options, such as immobilized yeast, cream yeast preparation, active frozen yeast and compressed yeast. These choices, along with the matter of how the yeast is added, are largely practical; I am focusing here on the potential effects on the style and flavour profile of the wine.

Ambient yeasts

So, noting that 'choosing the right yeast for your wine can greatly impact the taste, aroma and overall quality of the wine', what species of yeast to choose? Usually, the most common non-*Saccharomyces* yeasts in the air are genera such as *Kloeckera*, *Candida* and *Pichia*, those that give the flavour nuances often associated with 'wild-fermented' wines. Other genera may add particular characteristics, such as *Lachancea* promoting the relatively smooth lactic acid and some *Torulaspora* species producing lower levels of acetic acid. *Brettanomyces*, 'Brett', will be known to many wine enthusiasts. Its name, 'British fungus', comes from its discovery at the Carlsberg brewery in Denmark while investigating flaws in British beers. In wine too, its presence is normally seen as a fault, though for some its earthy, farmyard notes add to the taste of some Pinot Noir wines.

There is much debate about the values of these yeasts in promoting wine individuality and whether their source is principally in the vineyard or the cellar. However, in practice many commercial winemakers will not want to gamble with the unpredictability of truly 'wild' yeasts and would prefer to have microbiological control of the fermentation by using instead a commercially produced yeast.

One cunning strategy is to use non-*Saccharomyces* species that have been isolated and then cultivated under controlled conditions in a laboratory. Also, enterprising winemakers who are concerned about a perceived growing similarity in the world's wines may consider employing newly developed, *cross-species* hybrids, touted as potentially contributing new and different notes, 'additional diversity and complexity in the flavour profile'.

The virtuous yeast

Despite all this, however, far and away the dominant species used for wine is *Saccharomyces cerevisiae*. It's uniquely tolerant of alcohol, which means that as fermentation proceeds it soon outcompetes rival species, and it has long been recognized as producing wines of quality. This isn't the end of decision making for the winemaker, though, seeing as more than 700 different strains (sub-species) have been recorded. Most of these are of largely academic interest, but a wide range with different properties are available for purchase. So, which strain is appropriate for the kind of wine I'm aiming at?

Lalvin's 71B and K1-V1116 both produce a high level of esters, giving 'fresh floral aromas', an 'aromatic profile well appreciated by tasters'. Wyeast's 4267 produces a 'distinctive intense berry, graham cracker nose and jammy, rich, very smooth complex profile', 4242 gives 'bready aromas with vanilla notes', whereas 4028 encourages a 'Beaujolais-type fruitiness'. Powerful, elegant wines with 'non-aggressive tannins' are said to be produced by Laffort's Zymaflore FX10.

Or perhaps I should think more about mouthfeel, and consider Laffort's Zymaflore F15, which gives high glycerol production (ie, weight on the tongue). Anchor Oenology WE 372 gives high levels of the tart, salty-tasting succinic acid that some commentators have related to wine minerality.

All this will look very technical; the yeast names are awkward and unmemorable. In these ways it's very similar to the rootstock situation: presumably not the kind of thing that an average magazine buyer, say, wants to read about. Perhaps that's why we never see yeasts mentioned in the popular wine press, even though they are clearly relevant to wine taste.

The logic seems plain. Winemakers are choosing yeasts to help them achieve the kind of wine they're striving for, so coming at it the other way round, a wine we are tasting must owe something of its character to the yeast that was used. And after all, it's the yeast, ie, the fermentation, that produces the majority of the compounds that give wine its flavour: it's why wine doesn't taste like grape juice.

A work unacknowledged

In front of me is the current issue of a popular wine magazine; it summarizes the results of what is apparently the world's biggest wine competition and includes detailed tasting notes on the winners. There are 25 pages of comprehensive wine descriptions, with numerous mentions of how the taste of a particular wine has been influenced by some aspect of the climate or of the vineyard soil, and in some cases by particular things like nearby forests, proximity to a lake, or the casks used for maturation. But nowhere do I see a mention of the yeast that was used to ferment the wine.

Incidentally, all this contrasts very strikingly with the world of beer. Yeast is the defining characteristic of many beers. For instance, any beer enthusiast will know that ales are produced by top-fermenting yeasts while lagers are the result of those that ferment at the bottom; that the contrasts between, say, a Weissbier, a Dubbel or a Lambic derive from the different yeast strains involved. A number of classic English breweries boast of their pure, unique strains of yeast, some of them decades or even centuries old (with reference samples being held, along with 600+ others, at England's National Collection of Yeast Cultures, in Norwich).

But yeast in everyday wine writings on wine? Well, it seems *persona non grata*. Or, if you will, it's another 'elephant in the room'.

4 All in the mind

You're sipping a glass of wine and enjoying its depth and rich, velvety texture. You then have another glass of the wine, but now it seems different, decidedly edgy with a sharp raciness that you just didn't get before. It's the same wine, same bottle, so why the difference? Well, might it be due to a change in the music you have playing in the background?

Most wine enthusiasts will know about extraneous factors that can affect taste perceptions: best known probably is what we already know about a wine, its pedigree, cost, critics' ratings, etc. It's all in the mind. So the 'elephant' of this fourth section isn't about a factor that is quietly impacting the wine itself but one – psychology – that influences how we perceive it. Here, I look briefly at a few examples, working outwards from the wine in the glass.

Just one more glass

A primary driver of what we expect from a wine is its colour. It's been demonstrated in numerous studies, and some have shown how easy it is to fool even wine experts simply by deliberately miscolouring a wine. Colour trumps what our other senses are telling us.

The shape of a wine glass can be important. It affects the aeration of a wine and how it flows into the mouth, though the amounts are pretty much below our sensitivities. Rather, the effect is psychological. It's particularly influential when a taster is using a glass designed for a specific wine, usually a particular varietal, and knows that is the wine being tasted. (If the taster doesn't know such things, though, it seems the glass shape doesn't make much difference.)

Glassware influences our view of the status of a wine, for instance thin walls and stems tend to signal prestige, as does the wine bottle. There, information on the label is, of course, paramount, but in addition, shape, colour, label design can all affect our taste perceptions, and greater bottle weight suggests quality (and hence a dilemma for producers with environmental concerns). One analysis compared bottle weight and wine price and found that on average as the shelf price increased by £1, the bottle weight increased eight grams. Put another way, for every additional £1 you pay you get an extra eight grams of glass!

What did you have for breakfast?

How consistent is your tasting? Not very, judging by a plethora of research reports. How well did you sleep last night? Any anxieties, allergies, medications, supplements? Apparently, all these things can affect our taste faculties. And it seems our receptors for taste and smell decline with age, especially after 70…

What about the last time you ate? Apparently, taste recognition thresholds are highest immediately after a meal and then fade over a few hours. It does, though, depend on the food. Spicy enchiladas have a greater effect than a bowl of milky oatmeal, for instance, and because of their different satiating powers, meatballs have a more significant effect than macaroni. And on it goes.

Modern genetic technology is revealing all sorts of dispositions to certain flavours that may be built into our DNA. If you happen to have the snappily named TAS2R38 gene in your DNA then you are likely to be more sensitive to the alcohol in wine. This is the same gene that has acquired some fame (notoriety?) for determining our enjoyment or otherwise of eating Brussels sprouts. If you have that gene then sprouts will probably taste disgusting to you; if you don't then Brussels sprouts will taste, well, like Brussels sprouts.

It's a beautiful wine. And is that Beethoven I hear?

There's a host of issues to do with the setting in which you might be tasting a wine, the ambience if you like. A wine will taste very different at a funky party than it would at a funeral. That enchanting local wine you discovered while away on holiday famously becomes quite different when you drink it back at home. Also influential are your companions at the time, the size of the room, and the lighting (in one analysis, participants were willing to pay more than a euro more for a bottle of Riesling when it was offered in red rather than green light).

Are there flowers in the room? Even the venerable Mrs Beeton was aware that flowers influenced the taste of food. Each of the menus given in her 1861 *Book of Household Management* included a 'vase of flowers' to be present, carefully chosen so that the food flavours and colours 'contrast nicely'. Much more recently, a study showed that wines were perceived as being more 'delicate' when tasted in the presence of a similarly delicate flower arrangement (based on colour, texture, size, etc) as compared to a 'robust' arrangement or no flowers at all.

And then there's that matter of music. Most people rate wine more highly (and are prepared to spend more money) if any background music is classical rather than some latest pop hit. In one study, hearing bass notes resulted in wines being rated as significantly fuller bodied than when tasted in silence or with a higher-frequency sound. Smoothness and depth are enhanced by low-frequency, harmonious major chords whereas acidity and freshness are heightened by dissonant, high-pitched music played staccato. (All this may sound somewhat implausible but I can vouch for it, having been involved in one of the trials myself.) So perhaps that wine you were drinking at the start of this section was accompanied first by some 'legato' cello melody in C major but later by brisk modern brass and percussion sounds?

Star quality

A few years ago there was a buzz in the media about wines tasting differently on different days, according to whether it was supposed to be a so-called root day, a fruit day, a leaf day or a flower day. The idea was that on fruit and flower days those features were highlighted in a wine, whereas leaf days brought out vegetal aromas; on root days wines were relatively dumb.

So what determines the kind of day? It all depends (if I understand it correctly) on where the Moon is in the celestial zodiac. As one disciple puts it, 'the Moon opens a gateway from each constellation for a particular element force (*sic*). And that element enhances a particular aspect: earth/root, water/leaf, air/flower and fire/fruit so that those qualities in a wine are brought out'. Thus, somehow, a flower day occurs when the Moon is traversing the constellation of Aquarius whereas fruit days are when the Moon is crossing Libra.

This is all very alien to science but it was popular and great fun, until the whole idea was torpedoed by a group of cognitive psychologists. Following a series of rigorously designed taste tests their conclusion essentially was that the scheme didn't work unless you already knew what kind of day it was supposed to be. If you did, then a kind of placebo effect kicked in: you sensed what was expected. The notion still has its believers, but belief it is. We have seen plenty of reasons why a wine can taste differently on different days; it takes faith to ascribe the effect to the distant cosmos.

Placebo power

We've come a long way in this essay from a wine glass in your hand to constellations hundreds of light years away but the journey has given us glimpses of the multitude of things that can alter our perceptions of a wine. In view of them all, it's perhaps surprising that there's any consensus at all on wine tastes!

The thread through all the examples is that it seems there's an unwritten assumption that interesting though all these psychological phenomena are, they're academic: they don't actually apply in real tasting notes. The implication seems to be that a report on how a wine tastes will tell us – irrespective of the circumstances – what we can expect to taste. And given the suggestive power of the placebo effect, it's likely that's what we will taste.

16. Winescape GB

Winescape GB

In 1972 Steven Spurrier was trying to import five cases of English wine for the British Embassy in Paris when he got a phone call from French Customs and Excise. This shipment cannot be cleared, he was informed, because 'There is no such thing as English wine'. Winegrowing in Britain has always had challenges.

Just now, though, it's booming. In the first edition of *The World Atlas of Wine,* published in 1971, Hugh Johnson's masterly survey of the world's vineyard regions has at the end of the book a single page, looking rather like an afterthought, entitled 'England and Wales'. It lists 24 vineyards. In *Sotheby's World Wine Encyclopedia*, published 17 years later, the number had swelled to 142 (though that book's author went on to declare that 'over the next 10 years the number of wineries will reduce… wine in Great Britain is destined to remain a cottage industry').

As I write this in 2023, there are about a thousand commercial vineyards in England and Wales, including some major players and, judging by the growth in recent years, by the time this appears in print another hundred or so will have come into being.

This is just the most recent stage in a long and erratic history. The story begins 400,000 years ago, long before the advent of modern humans, when *Vitis vinifera* was growing naturally in the Britain: seeds have been found in sediments from those times. Those indigenous vines, however, were wiped out by the subsequent Ice Age. So who reintroduced grapevines to the Britain and who first made wine here? Very probably the Romans, but it's hard to prove. Archaeological digs are great for burial tombs, stone walls and the like, but finding evidence of indigenous farming is more of a challenge.

And the historical literature can be interpreted in different ways. In 90 AD Emperor Domitian outlawed all viticulture north of the Alps, without mentioning particular places; in 280 AD Emperor Probus lifted the ban – and specified inclusion of the British Isles. Therefore, the reasoning goes, vineyards must have existed there before Domitian's ban. Not at all, a later writer snorted: Emperor Probus – well aware of his legionnaires' opinion of Britain's climate – was simply showing his sense of humour. Daines Barrington (probably best known through his correspondence with the great English naturalist Gilbert White of Selborne) maintained that wine grapes were never grown in England in Roman times. It was all, he declared, a dreamy myth deriving from careless translation of Pliny's Latin.

Without doubt a high spot in British vineyard history was the Middle Ages, both with the flourishing of the monasteries, most of which had vines of some kind, and with land being doled out to the French compatriots of William of Normandy in the wake of his ruthless conquest. Things got tricky after the Middle Ages though, when wines from the nearby continent became more readily available.

Even so, some individuals tried to encourage English viticulture. John Rose, in his tellingly titled *The English Vineyard Vindicated* (1666) and William Hughes (in *The Compleat Vineyard: or a most excellent way for the planting of vines* – 1665) maintained that the real problem with growing vines in England wasn't climate or soil but *indifference*. 'I have just cause to accuse the extream negligence and blockish ignorance of our people,' wrote Hughes. Earlier, William Camden, the great English mapmaker and historian, had remarked that 'there are not Vineyards in England at present' because of 'negligence and sloth'.

Nevertheless, those early authors enthused about the potential for English wine and among other things insisted that grapevines could easily cope with the wide range of soils and terrain in Britain. Hughes remarked that vines 'prosper well in the stiff and fat ground, as in the soft and mellow ground; and oftentimes in the leany and lean ground, as in the fat and foggy; yea and in many places in the very Rocks, and gravily ground'. This is all just as well, because Britain displays, for its size, an astonishing variety of geology.

A geological melange

In this chapter I will scan the British vineyard scene as it is today in the context of this varied geology. I will set things in the context of geological age, though I must emphasize that this isn't itself relevant to the growing vines, let alone the wines. It's just an organizational framework, as we take an imaginary journey from northwest to southeast.

It's a handy general rule that the British bedrock decreases in age from the very oldest in the far northwest of Scotland through to the youngest in East Anglia.

Because of this geology, to the northwest is 'upland' Britain, with tough, older rocks tending to make upstanding land with poor, thin soils. It's always been harder to eke out a living there and communications were always difficult, all a contrast with the generally flatter, fertile land to the southeast, developed on softer, younger rocks. So, coupled with the overall rise southwards in temperatures and growing-season length, it's hardly surprising that the concentration of vineyards increases towards the southeast. Discussions of the British viticultural scene usually focus on the latter region but here I will include some of the lesser-known efforts further northwest, on older bedrock and with more challenging conditions.

Môn Mam Cymru

In 1191 Geraldus called the island of Anglesey 'Mother of Wales', or Môn. It displays the largest expanse of ancient rocks, so-called Precambrian, in the whole of England and Wales. And there are vineyards!

If you drive over to the island on the A5, over Telford's pioneering 1825 suspension bridge, you soon reach the town that rejoices in its 57-letter Welsh name, 43 of which are consonants. A left turn there and after 14 kilometres (9 miles) you reach Tŷ Croes vineyard, right behind the 12th-century Llangeinwen church. The bedrock here is around 600 million years old and very unusual (it's technically something called a blueschist). The soils, though, as is usual, are vastly younger, here deriving from the deposits of the geologically recent ice age. Most are clayey, but there are sandier, better-drained patches in the vineyard.

A few patches of rocks of similar age or even older occur in England – the Malvern Hills and Charnwood Forest are examples – though the nearest vineyards to them are on surrounding younger geology. But such exhumed areas provide valuable windows through the younger, covering strata to what lies beneath the whole of England and Wales, the geological 'basement' (analogous to the 'crystalline basement' I discuss in Chapter 8 on the Loire).

Cambria

Back over the Menai Strait and south of the town of Caernarfon, with its UNESCO-designated massive medieval castle and town walls, is the Nantlle

Valley and the Pant Du vineyard. This is slate country, part of another UNESCO World Heritage site: 'The Slate Landscape of Northwest Wales.' The soils here are reasonably well drained, particularly on sloping sites such as at Pant Du. Quarries hereabouts roofed much of southern Britain in Victorian times, indeed other parts of the world also. (I have seen wineries in South Australia's Barrossa Valley roofed with slate from here – Moorooroo and Jenke, for example.)

Slate is all around at Pant Du, from where it continues over the hill north of the vineyard into the next valley, to the area around Llyn Padarn. There, in 1831, two pioneering young men were visiting to examine the rocks. One of them – a student 'mad about geology' – had to leave the area after a few days to return home to Shrewsbury, where he found a letter waiting for him. It was an invitation to join an upcoming ocean voyage, on a ship called HMS *Beagle*. Mr Darwin agreed to go.

The older of the two was Adam Sedgwick, professor of geology at Cambridge. While at Llyn Padarn he decided that the rocks there were so different that they needed their own name. He deduced that they formed during a particular interval of geological time, which he called 'Cambrian'. The name comes from the Roman name for Wales, Cambria, in turn based on the Welsh name Cymru (roughly 'cum-ri'). Today, geologists around the world use the word for rocks formed around 500 million years ago. (Geologists visiting me here in mid-Wales find it hilarious that the local newspaper is called the *Cambrian News*.) It appears on wine labels and in promotional materials at places like Spain's Castilla y Leon and Australia's Heathcote: 'the ancient Cambrian soils add depth and complexity' and the like.

Producing roof slates is notoriously wasteful and Pant Du makes good use of the material discarded from nearby quarries, using it as a mulch around vine roots and melting it at over 1,500°C (2,700°F) to make special bottles. That may sound radical but, after all, standard glass is made of melted sand extracted from the ground.

Celtic names echo around the world

Stretching south down through Wales is bedrock formed during the two periods that followed the Cambrian, also named after things Welsh. The Ordovician period, around 450 million years ago, was named after a pre-Roman Celtic tribe that peopled northern Wales, and the Silurian, around 430 million years ago, after the Silures folk of southeast Wales. The region was sea in those times and the shales and mudstones that formed tend to give today's thin soils, which are clayey and poorly drained.

There are a few small but thriving vineyards based on this bedrock: Llaethliw, near Aberaeron, for example, and Whinyard Rocks at New Radnor. Pentreheyling near Church Stoke is sited on old lake deposits covering the Silurian bedrock. All this landscape is, in a word, grey (often not unlike the weather!). The vineyard soils are grey, as are any stones in them; buildings and field walls of the local stone are grey. But to the east of all this, everything changes. The land becomes decidedly ruddy.

The place where this change was first described is in the Wye Valley, just south of the little village of Erwood. Sir Roderick Impey Murchison, landed gentleman and retired soldier, spent much of his time shooting birds and hunting animals – by one account he was 'the most celebrated foxhunter in England'. But in the summer of 1831 he was 'geologizing'. Murchison visited the Wye Valley and at Erwood saw red rocks and soils to the east, across the river, but at his feet quite different, grey rocks. He named them Silurian. Today there are vineyards around the world on bedrock formed in Silurian times – in Spain's Penedès, for instance, and Alsace's Kastelberg. A wine from Canberra, Australia, is called simply 'The Silurian'.

Change to red

Towards the end of the Silurian period, in this part of what is now southern Britain, the sea receded and the resulting land became arid. Red sands and silts accumulated in the desert conditions and eventually became ruddy rocks such as those that Murchison saw across the Wye. They are still referred to by the time-honoured name of Old Red Sandstone. (And yes, there is a New one, as we'll see shortly.) The peak of their development was during the geological period that followed the Silurian: the Devonian. We have already met the geologist who in 1839 coined this name after his work in Devon: Adam Sedgwick.

In contrast to the grey, hilly country to the northwest, the Devonian Old Red Sandstone underlies thick, fertile, well-drained soils on generally flat land – the orchard and cattle country of Hereford, for instance. There are a number of vineyards on Old Red Sandstone, such as Llanerch, between Cardiff and Bridgend, the Sugar Loaf at Abergavenny, and nearby White Castle. The wines from Ancre Hill at Monmouth, according to their publicity, 'express the ancient

OPPOSITE: Views of the Pant Du vineyard, in the slate-rich Nantlle Valley, Gwynedd, North Wales. Out to the west lies Caernarfon Bay and to the east Yr Wyddfa (Mount Snowdon).

mudstone and sandstone soils, which give the wines their character and natural flavours'. (Ancre Hill wine has been served at a meeting of G20 world leaders.)

In July 1798, William Wordsworth returned to a place that had haunted him since a previous visit: the Wye Valley around Tintern Abbey. 'Once again do I behold these steep and lofty cliffs,' he wrote, and steep the valley sides are at Tintern. They are firmly in Old Red Sandstone country, by canoe more than 100 kilometres (60 miles) down the snaking river from Erwood, and here the Wye has sharply incised into the bedrock. High on a steep slope above the river, looking towards the storied abbey, is a vineyard that can lay claim to being the first commercial Welsh operation of the modern era.

There had been a notable vineyard at Castell Coch, just north of Cardiff, but it closed in 1920. That endeavour was something of a self-indulgence of the exceedingly rich Marquess of Bute, who fancied a vineyard next to the turreted fairytale castle he had built. In 1873 the Marquess had sent his head gardener (who spoke no French) to Champagne and the Médoc, to learn about wine. He came back with a few white-grape vines, apparently of a variety called Miel Blanc, but the gardener seems to have been enamoured with the idea of producing red wines in Wales. Understandable, perhaps, insofar as he spent most of his visit in the Médoc, including observing the vintages at châteaux Latour, Lafite and Margaux!

However, it was Gamay that he planted at Castell Coch and by 1893 all the production was red. It was always a challenge. Some years the crop failed completely and even in good years the grapes needed substantial additions of sugar. Then, sadly, in the years of World War I, all that sugar just wasn't available, and the effort had to come to an end. So in 1920, the 63,000 vines were grubbed up. And for the following 59 years there was no such thing as Welsh wine. (And precious little English wine, for that matter.)

Thus the planting in 1979 of Seyval Blanc and Müller-Thurgau vines near Tintern Abbey heralded a new era in Welsh wine. The Tintern Parva vineyard has been something of a beacon ever since, though now joined by more than 30 other Welsh producers. The bedrock at the Tintern vineyard 360 million years ago was a riverbed in the Devonian desert, but now yields well-drained sandy soils. This same Old Red Sandstone formation continues west along the hillside from the vineyard as far as the intriguingly named Barbadoes Hill; it was from there that stone was once quarried and carried down to the river, to build the celebrated abbey.

Time moves on

The red desert lands of Devonian times fringed a major ocean to the south such that the Devonian age rocks we see in southwest England (the bedrock in between comprises younger strata) are quite different in nature. Many are grey sandstones and mudstones, some now compressed to slate. The rocks are very visible around Tintagel, supposed site of King Arthur's castle, and at the huge Delabole quarry. Vineyards such as Polgoon, Bosue, Looe Valley and the long-established Camel Valley – all in Cornwall – involve soils at least partly derived from slate; Sandridge Barton (previously Sharpham), in Devon, has volcanic bedrock.

A hundred million years after all this, masses of granite were forming deep below the ground of the time, but have since been excavated by erosion. They now make renowned landscape features of Devon and Cornwall such as Dartmoor, Exmoor and Land's End; Knightor vineyard is sited on soils derived from the Saint Austell granite.

The interval that followed Devonian times is known as the Carboniferous. It's a misleading name in that the swampy conditions that produced the coal-bearing formations indicated by the name existed only in the last of the three divisions of this period. In the early times, warm shallow waters allowed coral reefs to flourish, which led to the formation of limestone. This Carboniferous Limestone is a tough rock, dense and hard, and makes land that in places is almost devoid of vegetation. It forms distinctive upland scenery, as in the Mendips, the Clwydian Hills, the South Wales cave country and, perhaps most famously, the Yorkshire Dales.

These limestone times were followed by the development of a large delta in which sands and muds accumulated. The resulting rocks – from the middle division of the Carboniferous – are collectively known by the evocative name of Millstone Grit. Many a Pennine village has its name proudly affixed to a wheel of Millstone Grit, the stone once used hereabouts to grind corn. There are poems and novels about the rock; when Emily Brontë had Catherine say, 'My love for Heathcliff resembles the eternal rocks beneath', it was a reference to the obdurate Millstone Grit that underlies the high moors of Brontë country. Obviously, the hilly country that these rocks produce, together with the northerly latitude, is not conducive to vineyards. Nevertheless, there are some.

Sheveling Wine Estate, which manages to grow Seyval Blanc at an altitude of nearly 300 metres (1,000 feet), overlooks the town of Holmfirth, West Yorkshire. All around are the coarse sandstone strata of the Millstone Grit, giving well-drained, buff-coloured sandy soils. Whirlow Hall vineyard, near Sheffield, is sited on similar bedrock. Amber Valley vineyard, just outside the Peak District National Park, calls its Solaris wine simply 'Millstone'.

The end of the Carboniferous period was heralded by a return to arid, desert landscapes during two periods called the Permian and the Triassic, often linked together. And here we have the rocks of the New Red Sandstone. It's a fine building stone – think of the cathedrals in, for example, Lichfield, Exeter, Coventry, Liverpool – and it yields warm, fertile soils that underlie a host of vineyards, stretching from near York in the north to Teignmouth in the south. Two examples, both well established and long providing winemaking facilities for other vineyards, are Halfpenny Green and Three Choirs.

The charmingly named Halfpenny Green (pronounced, as any Brits older than about 60 will know, 'haypenny') was first planted in 1983 and continues to grow, being handily located in Staffordshire just west of Dudley, close to the populous areas of the West Midlands. There are patches of glacial boulder clay in the vineyard but the bedrock is sandstone of Triassic age, actually the so-called Sherwood Sandstone, a vital aquifer for Central England.

In 1715, the Chancellor of Hereford Cathedral proposed to his members and those of the nearby cathedrals of Gloucester and Worcester (itself built of New Red Sandstone) a 'friendly proposal, between lovers of harmony and brethren of the correspondent choirs, to commence an anniversary visit'. The resulting annual celebration, now billed as 'the oldest non-competitive classical music festival in the world', is invariably known simply as the 'Three Choirs'. In 1973, vines were planted just north of Newent in Gloucestershire, and was later named the Three Choirs Vineyard to reflect the owner's connections with the choral festival. It's now one of the largest and most respected vineyards in the UK. Like Halfpenny Green, the plots are on gentle south-facing slopes of Triassic sandstone, with rosy, sandy soils, pebbly in places.

Scarp and vale country

Mary Anning was only 11 when she found in the cliffs at Dorset's Lyme Regis her first complete ichthyosaur. And, in a pattern that was to beleaguer much of her life, the deprived child sold it for a mere £23 to a gentleman of wealth (in this case, the squire of what is now Royal Sandringham, in Norfolk). Ichthyosaurs lived in times known as the Jurassic, perhaps the best-known geological period of all, thanks to the Mary Anning story, together with the Dorset cliffs now being part of a UNESCO World Heritage site; and, of course, to movies and dinosaurs.

The Jurassic saw a return of warm, life-filled seas, fluctuating in depth such that over time sequences of mudstones, sandstones and limestones were produced. They make something of a layer cake, tilted by later Earth forces towards the southeast, with the strata becoming eroded differentially to give the scarp and vale scenery that characterizes much of central southern England.

If you drive southeast on the A417 out of Gloucester, across the flat clayey and silty rocks formed in earliest Jurassic times, you are suddenly faced with the steep climb up the Crickley Hill escarpment, passing upwards through sandstone and limestone strata to the major limestone at the top that dominates the Cotswolds plateau. It yields the cream-coloured 'Cotswold stone' that characterizes so many picture-postcard villages in the region.

This same arrangement is reflected at Gloucestershire's Woodchester Valley vineyard. As you weave up to the winery from the clays at the bottom of the narrow valley from Nailsworth, you pass various limestone strata, then reach the main limestone at the top. The stone and clay contents of the soils vary: according to the producers, Pinot Noir and Chardonnay vines do best on the stony limestone whereas the thicker, more clayey soils are more suited to Bacchus. And in a nice illustration of the fluidity in the present English wine scene and of its promise, a Woodchester wine was ranked in a 2023 international blind tasting as among the very best in the world. A Bacchus perhaps, or Solaris? It was a Sauvignon Blanc…

Strata of Jurassic age stretch all the way from Redcar on the northeast coast down through central England to Lyme Regis, with vineyards spread across the varying rocks and soils. I notice that Hugh Johnson's handsome 'Map of the English Vineyards', published in 1989, stops at the latitude of Nottingham and Stoke-on-Trent; today, vines are being planted in the far north of England (so not on Jurassic bedrock) and even in Scotland. Right now, though, it seems Ryedale Vineyard, near Howsham, north of York, is the northernmost commercial, producing operation in the UK. The plots there are partly on sandstones and shales formed in the large estuary that was here during middle Jurassic times, about 160 million years ago, and partly on a limestone rather like that so dominant in the Cotswolds.

The land at Ryedale is pretty flat but just five kilometres (three miles) to the east things abruptly change: the road climbs to the distinctive upland terrain of the Yorkshire Wolds, David Hockney country. As you climb the escarpment, picturesquely called Leavening Brow, you pass through Kimmeridgian clays from the geological age well known to wine enthusiasts, then greenish iron-rich sandstones and finally dazzling white chalk from the Cretaceous age.

Chalk and seas

Cretaceous bedrock formed between 145 and 66 million years ago, and today makes the next swath to the southeast of the Jurassic. It extends southward from the Yorkshire Wolds down through the Lincolnshire Wolds and central East Anglia, then swings southwest to the Chiltern Hills and repeats in a major upwarp – the Weald – to give the North and South Downs. And here we have reached the heartland of the modern revival in British viticulture.

Warm, shallow seas continued during Cretaceous times, though with depths and other conditions continually changing. The early part, about 110 million years ago, saw sands, silts and clays being deposited, and an unusual mineral forming on the sea floor. Decay of the silica micro-organisms that were flourishing in upwelling, poorly circulating nutrient-rich waters prompted formation of the iron-phosphorus mineral glauconite. It's greenish in colour and the sandstone rock that eventually resulted is known as greensand.

Some producers, Ambriel for instance, maintain that the good drainage and well-balanced nutrition provided by soils deriving from greensand are perfect for vines. Others disagree. In later Cretaceous times, the sea conditions enabled the flourishing of tiny marine organisms made of calcium carbonate, and especially one that habitually shed the little platelets that clad its shell, properly called coccoliths, and hence the production of the rock called chalk. People often call vaguely white, powdery limestones 'chalk' but to a geologist the defining characteristic of true chalk is the dominance of these coccoliths.

And some producers contend, Rathfinny for example, and Hambledon, that this is the ideal bedrock, at least for the production of sparkling wines. In the days when the idea of decent English sparkling wine was fighting for credibility, much was made by some writers and producers that the North and South Downs were made of the same unique rock that underlies the Champagne region. 'Our vineyards are anchored in the iconic chalk that stretches from southern England to the French Champagne region, without doubt the most desirable for producing the world's finest wines', that kind of thing. Some journalists became almost fixated on the idea.

Certainly, chalk's water-retaining properties and restrained nutrient provision make it an excellent substrate for viticulture (though iron deficiency can be a problem, which is not the case with greensand). Actually though, less than a third of the Champagne region is based on chalk and the rock is variable in detail, both there and in England. Moreover, it's hardly unique: the formation extends intermittently eastward as far as the hills of Crimea and Kazakhstan; those knolls around Austin, Texas, and the oil-bearing rocks below Dallas – they're all chalk.

Some major producers have sites on both greensand and chalk: Nyetimber, for example, and Ridgeview. But in this region of greatly varying terrain, together with proximity or otherwise to the caprice of the English Channel, it's the site that's pivotal, more than the chalk or greensand involved. In fact, some top vineyards are on other kinds of bedrock altogether, such as sandstone (Bolney, for instance). Gusbourne vaunts the 'roundness, fullness and body' that clay brings to its wines: 'it can outweigh chalk in England', maintains the head winemaker. And now there are signs that the British wine spotlight may be moving into distinctly clayey country.

Go east

Around 50 million years ago, in so-called Eocene times, the area that is now eastern England was a sheltered, quiet sea in which thick quantities of mud accumulated. Some of it hardened to become what geologists call the London Clay. And crucial it is for the region.

For example, it's a fine material for making bricks, the basis of much of London's housing, and it lends itself to tunnelling, thus allowing the rapid expansion of the London Underground. The Tube lines are shallower north of the Thames because the London Clay is shallower there. It's also thicker, reaching its maximum around the river Crouch in Essex – the newest focus of the broadening English wine scene. This is just about the warmest part of England and has the driest air. Moreover, the flat clay land lends itself to mechanization, an important commercial consideration with Brexit having curtailed the availability of manual fruit pickers.

Californian money and interest from, rumour has it, 'five or six big names in Burgundy' is being directed at the Crouch Valley. One winemaker has quipped that the quantity of grapes being grown in the valley is 'increasing hour by hour', and a major new contract winery is being built, specifically to make ultra-

premium still wines. A commentator recently wrote of 'the vinous epicentre just to the north of the meandering river Crouch, England's Côte-d'Or, now producing some of England's best still wines'.

And from here?

A couple of recent academic studies of land conditions and climate trends in England and Wales have reported tremendous viticultural opportunities. The analyses revealed promise in areas such as Suffolk, the East Midlands, the Severn Valley, southwest England and South Wales. And the largest temperature increases were forecast for parts of Essex such as the Crouch Valley: not only was Pinot Noir predicted to ripen reliably there, but possibly Sauvignon Blanc, Riesling and Semillon.

Obviously all this is very encouraging for the British wine industry and it should make a fittingly positive note for me to end this chapter. Perhaps it finally signals the end to all those difficulties of the past? Unfortunately, however, I have

to sound a cautionary note. The rising warmth that is so opportune for helping ripen grapes in these more marginal climates also triggers problems – late frosts, hailstorms, disease and wildfires, for instance.

And the ice caps are already melting at an alarming rate. Inevitably this means sea-level rise, and even conservative estimates have an increase of two metres (six feet) in the next hundred years or so. It's not those viticulturally challenged areas we visited first, towards the northwest of Britain, that will be affected by this but those flat, low-lying areas far to the southeast that we've just seen.

In other words, Winescape GB in a century's time is likely to look very different from now. There may be many more thriving vineyards all around the country, happily ripening currently exotic cultivars – but some of what were the most promising potential areas will by then be under the sea…

ABOVE: Vineyards such as Clayhill, New Hall and Danbury Ridge are thriving on the clay soils around the river Crouch, here viewed over South Woodham Ferrers, Essex, England.

LAPHROAIG
DISTILLED
1978
ISLAY MALT WHISKY

17. What about Whisky? What about Beer?

What about Whisky? What about Beer?

Wine is not the only drink where a role for geology is trumpeted. Japan's sake, for instance, is said to be heavily influenced by the geology of the brewing water. An example is the revered *Nada no miyamizu* spring water, responsible for the celebrated sakes of Nishinomiya. The water comes from the granites of Mount Rokkō, near Kobe, which give it just the right dissolved mineral balance for the region's robust, 'masculine' style of sake.

Whisky, both Scotch and American, and some brandies like to make something of geology, though just as with wine – I would argue – on scrutiny it's all a bit more subtle than many of the claims would have it. But here's a curious thing. There is a drink where geology was of pivotal importance historically and still today can have a direct effect on taste, and yet we very rarely hear about the connection. Beer. Through the role of water – and beer is mostly water – geology can play a leading role, but one that is wholly unsung. Here I take a quick look at how geology figures in these various drinks.

Brandy

Many spirits derive their flavour from particular ingredients – tequila from the blue agave plant, Calvados and applejack from apples, rum from sugar cane – or from additives, such as juniper berries and botanicals in gin. Geology isn't usually taken to be of much relevance. Neither is it with brandy (distilled wine) apart from in two regions in France, Armagnac and Cognac. In both cases, a role for geology is vigorously promoted, though it becomes less than convincing when looked at closely.

In 1909 the Armagnac region in Gascony was divided into three, on the basis of municipal boundaries. But subsequently the geology of each subdivision became increasingly mentioned – sands in the wetter Bas Armagnac to the west, clay and limestone in the central Ténarèze, and marl and limestone in the drier Haut Armagnac to the east – and these days it's a routine part of promotional literature. The geological changes are, however, very generalized and, because the Armagnac region is built from sediments shed south from the Pyrenees, in detail they are very intricate, reflecting the abrupt changes typical of alluvial fans. In other words, the boundaries of the Armagnac sub-regions simply don't correspond with the geological map.

But such mismatch is even more glaring in the Cognac region. It's a good story, though. In 1857 the president of the Geological Society of France, Henri Coquand (he of the three-way division of rocks into igneous, sedimentary and metamorphic, *see* Chapter 1), led a field trip to the Cognac district. Nothing was made of any brandy–geology connection during the serious work of the day, but that changed during the evening. Things concluded with '*un diner splendide*' at the eccentric Château de Maillberchie. Monsieur le Président gave the after-dinner address and, at least according to the late Professor Jake Hancock, he injected humour into proceedings by declaring to the satiated diners (they missed the last train back to Paris, so generous was the meal) that the quality of the region's brandy actually depended on the geology.

Coquand noted that the brandy quality increased in diminishing circles to a bull's eye centred on the town of Cognac itself and then breezily pronounced that this came about because of the chalkiness of the rocks also increasing inwards concentrically. And today that is the mantra of Cognac promotion. However, the chalkiness doesn't increase like that at all, as Coquand knew very well. The maps of Cognac quality and bedrock simply don't correspond. The yarn was, according to Hancock, wholly tongue-in-cheek and Coquand never intended it to be taken seriously. Little did he know…

Scotch whisky

The bulk of the world's whisky is manufactured in large plants, many of which produce a range of styles in one factory from bulk grain, adjusting the chemistry of the water supply as needed. Geology seems of little relevance. However, it is of potential influence in the 'handmade' characterful single products of relatively small distilleries, epitomized in Scotland and in Kentucky and Tennessee in the USA.

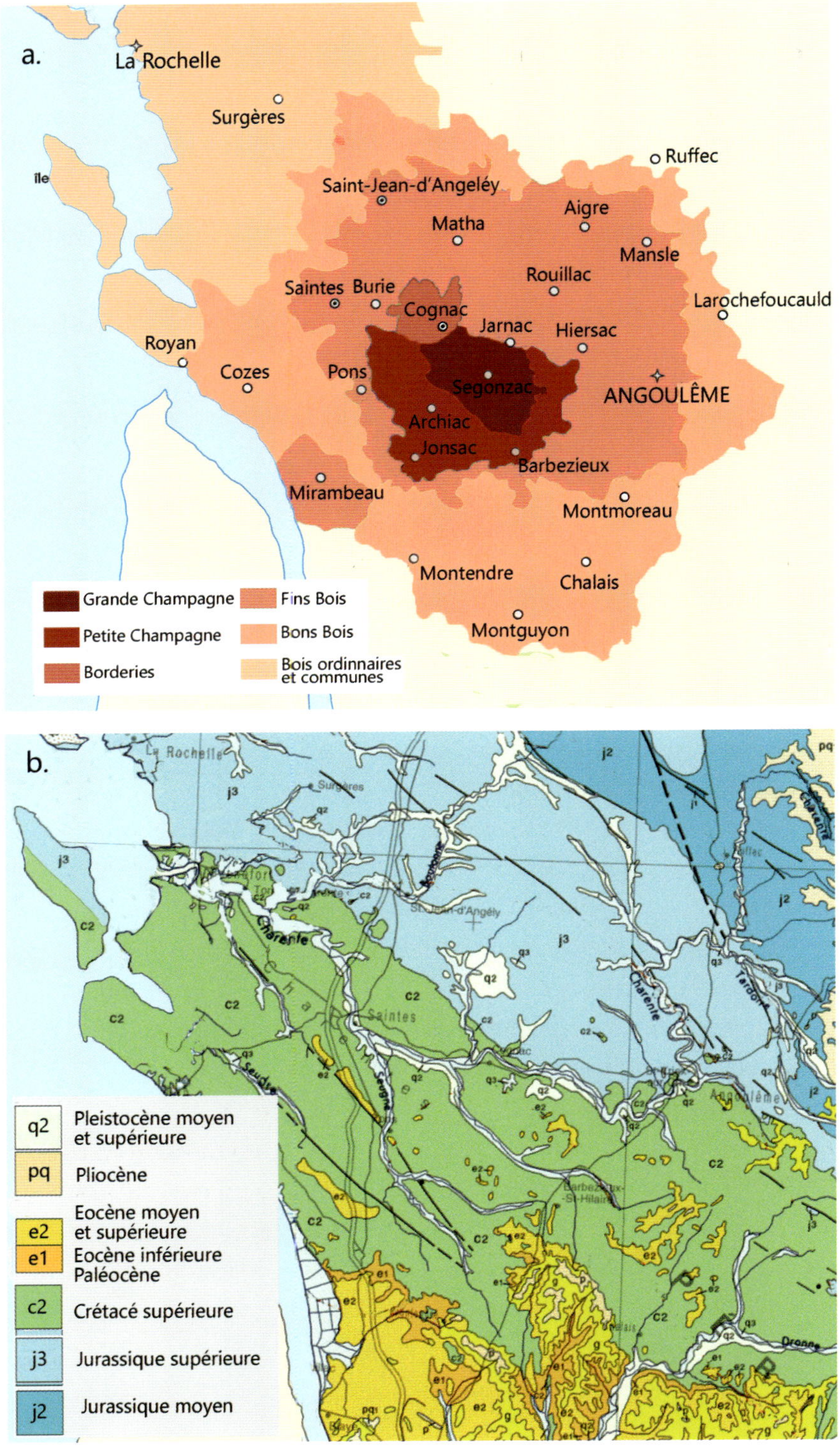
a.
La Rochelle
Surgères
île
Ruffec
Saint-Jean-d'Angeléy
Aigre
Matha
Mansle
Rouillac
Saintes
Burie
Cognac
Larochefoucauld
Jarnac
Hiersac
Royan
Cozes
Pons
Segonzac
ANGOULÊME
Archiac
Jonsac
Barbezieux
Mirambeau
Montmoreau
Montendre
Chalais
Montguyon
Grande Champagne
Petite Champagne
Borderies
Fins Bois
Bons Bois
Bois ordinaires et communes
b.
Charente
Saintes
q2 Pleistocène moyen et supérieure
pq Pliocène
e2 Eocène moyen et supérieure
e1 Eocène inférieure Paléocène
c2 Crétacé supérieure
j3 Jurassique supérieure
j2 Jurassique moyen

The finest of Scotch whiskies are the 'malts' produced by each of a hundred or so distilleries using only malted barley and relatively small, single batch, copper stills. The barley, perhaps having been permeated by peat smoke, is steeped in hot water ('mashed') and fermented with yeast, and the resulting liquor distilled. If any of these steps is significantly affected by water chemistry, as it is with the production of beer (*see* later), then the geology local to the distillery could be important for the whisky. The claim is certainly made.

'Soft water tumbling over granite rocks makes the whisky great'; 'the crystal purity of the waters, flowing down hillsides of ancient granite, gives the whisky a character that cannot be emulated.' Marketing of Scotch whisky likes to invoke unspoiled scenery, soft water and that much-loved rock, granite (*see* Chapter 4), a rock popularly associated with the Scottish Highlands. An image of granite mountains and crystal waters certainly makes for fine advertising copy.

However, not many distilleries use water that has been in contact with granite (Fettercairn, Aberlour, Arran and Glenfiddich are exceptions); some, Oban for instance, capitalize on the consistency offered by mains water.

More than half of Scotland's distilleries produce 'Speyside' whiskies, being located in the valley of the river Spey, from which some draw their water. Much of this river crosses rocks such as quartzite and schist, hence Speyside distilling waters do tend to be soft, but not really because of granite. In any case, the claim that this softness is the key to the character of Speyside whiskies is contradicted by the examples that use hard water, through having contacted calcareous rocks such as limestone. Examples are Glen Moray, Dailuaine and Cragganmore.

Some Speyside distilleries use groundwater, either from springs or from wells (Auchroisk, Macallan), and this can be distinctly hard. There are no Speyside whiskies more classic than Glenlivet, which uses water springing from limestone, and Strathisla, where, according to the distillery's own promotional material: 'The same water, with its calcium hardness, has been used since at least 1786.'

Whisky lovers see Speyside whiskies as different from those from elsewhere, such as that other pre-eminent area of quality malt production, the Isle of Islay in the southern Hebrides. Yet it is striking, as far as the role of geology is concerned, that most Islay distilleries are located on just the same kinds of rock that underlie

OPPOSITE: The upper map shows how the quality of Cognac increases concentrically inwards, usually claimed to be due to a matching increase in chalk; the lower map shows how in reality the geology fails to correspond.

much of Speyside, which yield the same ranges of soft and hard water. For example, the celebrated distilleries of Laphroaig, Lagavulin and Ardbeg use soft, surface water that flowed over quartzite, whereas Bunnahabhain uses hard water piped direct from springs in calcareous dolomite. Others of the island's whiskies are regarded as 'Islay' in style, but use a variety of different waters.

Another illustration that geology has no major role in characterizing whisky styles is in the 'Campbeltown' whiskies made by three distilleries (Glengyle, Glen Scotia and Springbank), each with a range of flavours. All have the same water source, a small loch located on similar bedrock to Speyside and Islay.

'Lowland' whiskies are also made from various water sources, as are the group loosely referred to as 'Highland'. Examples of the latter include Tullibardine, Aberfeldy, Blair Atholl and Glendronach – all inland and involving waters flowing over silicate bedrock. Coastal examples are different both in character and geology, though with no obvious connection. Some draw their water from sandstone, which has a significant calcareous content that makes the water hard. The distillery literature for Glenmorangie, for instance (located on sandstone), remarks that: 'The water is hard and 100 years old by the time it bubbles up in our Tarlogie Springs. Before modern filtration systems were installed, limescale had to be laboriously chipped away from the inside of our boilers.'

All this is clearly at odds with the marketing maxim of granite and soft water being key to Scotch whisky. It seems that any effect that geology has on whisky taste is swamped by other factors, such as the nature and quality of the starting grain, the degree to which it is peated, the architecture of the stills and the distilling methodology, and the nature and duration of the cask ageing.

However, there are other, subtle influences of geology. The distinctiveness of most Scotch whiskies lies in their peatiness and geology has some influence on the availability of peat. Consider the Isle of Islay, for example. It is no coincidence that this home of the most heavily peated whiskies has exceptionally thick deposits, and following tradition peat is still lavishly employed. The peat accumulated in extensive, flat, low-lying tracts, developed on a bedrock unusually soft for this part of the world, namely schists adjacent to and fractured by a regionally important fault zone.

The control of geology on landscape may have influenced the locations of the early distilleries, including some that went on to become famous (and legal). In the early days, taxation of whisky drove illicit production to clandestine locations, and a topography that allowed hidden sites and easy getaways must have been an

advantage. Southeast Islay is a case in point, where strips of tough igneous bedrock alternating with soft schist make a series of sharp ridges and narrow glens, and, just offshore, treacherous reefs, perfect for a quick getaway from the excise man.

American whiskey

It was only natural that the early American settlers, migrating westwards and claiming free land provided they planted corn (maize), brought with them distilling techniques from the mother countries but used this new kind of grain. And it could be argued that geology was key to the success of this new industry,

ABOVE: The Laphroaig distillery, Islay, Scotland, uses water from the Kilbride stream, soft through flowing over quartzite, and its maltings capitalize on Islay's rich deposits of peat.

through its providing copious supplies of good water and its yielding soils of a fertility ideal both for corn and the white oak, *Quercus alba*, for barrels.

Today, American whiskeys are still dominated by corn, but numerous other cereals are also used. Bourbon whiskey has to be made from at least 51 percent corn, and aged in new, charred barrels of American white oak; Tennessee whiskey goes through an extra charcoal filtering process. Today, Kentucky and Tennessee remain the whiskey heartland, but in these days of craft distilling it is a rare American state that has no distillery. The industry has many parallels with Scotch, and the claim that its secret lies in the geological influence on its water is even more glorified. But here it is for a reason opposite to soft water from granite: it is apparently distinguished by the pronounced hardness of the waters due to limestone. Certainly, as almost all the Kentucky and Tennessee distilleries are sited on limestones, the waters are uniformly hard: high in bicarbonate and with pH values as high as 7.5–8. But as with Scotch, the extent to which water itself affects flavour is moot. A whole range of tastes is produced from this hard water, and similar whiskeys are made elsewhere in the country with distinctly soft water.

ABOVE: Storage caves, mentioned in the report of the great Lewis and Clarke expedition to the West in 1805, at the Holladay Distillery, Missouri, USA, the oldest west of the Mississippi.

Steeping of the grains ('mashing') is ideally carried out between pH 5.2 and 5.5, so these naturally alkaline waters presented challenges to the pioneers. However, corn has micro-organisms such as lactobacilli capable of yielding sufficient lactic acid to reduce the pH values significantly. And in 1835, Inverness-born James Crow perfected a method of reusing spent yeast that had become acid in a previous batch to lower the alkalinity of the new mix. His 'Old Crow' whiskey, once America's biggest-selling brand, used this method, and today this 'sour mash' technique is used in all bourbon whiskeys.

The distilleries access the limestone waters differently. For example, Knob Creek and Woodford Reserve use spring water, as does Jack Daniel's (its location is even called Cave Springs). Maker's Mark draws its water from a spring-fed lake; Heaven Hill, Old Forester and Buffalo Trace use municipal water. New distilleries continue the lore of the limestone water in their advertising, and those elsewhere in the nation will almost certainly mention hard, limestone water in their advertising if it is at all nearby.

For example, what was until not long ago the only whiskey distillery west of the Mississippi, Holladay in Weston, Missouri, touts its limestone waters. As long ago (for this part of the world) as 1857, David Holladay recognized that the constant 12°C (54°F) 'bland water' flowing from a limestone spring would suit whiskey distilling, and that the nearby limestone caves (first mentioned in 1804 by Lewis and Clarke) would be ideal for barrel ageing. Much further west, Wyoming whiskey in Kirby uses 'Wyoming grains and water from a mile-deep limestone aquifer'.

But seeing as in the Appalachian–Midwest region limestone is widespread, other factors must have contributed to the pre-eminence of Kentucky and Tennessee, and to some extent geology was among them. The open ocean that was hereabouts in Ordovician times yielded more phosphorus than could be used by the marine organisms of the day, and the excess was precipitated leaving the phosphorus-bearing minerals that are common in today's bedrock. These now lead to soils unusually rich in phosphorus. In fact, in order to achieve a healthy calcium–phosphorus ratio, some farmers have to add lime to their soils, and this in turn leads to a pH range ideal for optimal nutrient uptake by cereals.

Not only does such fertility favour the growth of corn (and the famous 'bluegrass' of Kentucky – who would have thought that geology could engender a racehorse industry and a genre of music!), it helps account for the abundant indigenous white oaks that have always been used here for cask-ageing the whiskey. Legend has it that at one time a Reverend Elijah Craig used such barrels even though they had been partially burned in a cooperage fire, and by the time his whiskey had made it all the way down to New Orleans it had acquired a mellow, caramel taste and an attractive amber colour. This charring of American white oak is now central not only to modern bourbon but also indirectly to much Scotch whisky, which uses ex-bourbon barrels.

Beer

Beer is often seen as the simple quaffing drink of the masses, in contrast to the sophistication of wine. In fact, in many ways beer is the more complex drink of the two. One wit pointed out that if you stomp on grapes you're bound to get wine of one kind or another, whereas if you stomp on cereal grains you just get sore feet. Another suggested that when Jesus performed the miracle of turning water into wine, he avoided doing it to produce beer because that was too technically demanding.

Beer has an illustrious heritage. In the British Museum there's a clay tablet engraved with the oldest-known writing. It's from Mesopotamia and is over 5,000 years old – and it gives a recipe for beer. The royals of ancient times drank beer, well, they sipped it through a straw. Also in the British Museum is the beer straw used by the Queen of Ur in around 2600 BCE – made of gold and lapis lazuli, no less.

By the Middle Ages, in Europe brewing was everywhere, with many people believing that drinking plain water was actually dangerous. In England beer was drunk with every meal: the consumption was something like two pints every day, *per person*. The brewing was done by the woman of the house (rather like men always doing the barbecuing today), although the brewer was usually officially named as her husband. Some things change little.

Beer is mainly water, grains and hops. A variety of cereals is used in beer making these days, but most centre on barley. Areas suitable for barley growing are very largely dictated by overall climatic and agro-economic factors, but geological conditions can play a part. For example, in southeast England barley growing on clay soils performs differently from that on greensand (as with grapes, *see* Chapter 16).

Some hop-growers maintain that soil influences the aroma of their hops – in the 'minerally' character of Fuggles, for example – but the main role of geology with hops, again as with vines, is in providing free-draining soils with some access

TOP: Field of barley, near Eyemouth, Berwickshire, Scotland. BOTTOM: Hop flowers, which contain compounds used to stabilize and add bitterness to beers, in the author's own garden.

Burton Salts100g
Burton Water Salts
Contains: Papain, gypsum.
Add 1/3 oz per gallon just prior to boil.
Net Wt. 1 lb (454g)
Dublin Salts
Dublin water salts. For emulating Dublin water. Adding this pack to 5 gallons of distilled water will yield:
118 ppm calcium 5 ppm magnesium 25 ppm sodium 50 ppm sulfate 16 ppm cloride 200 ppm chalk
Dortmund Salts
Water salts for emulating Dortmund water. Add to 5 gallons of distilled water. This salt pack will add:
172 ppm calcium 23 ppm magnesium 72 ppm sodium 240 ppm sulfate 101 ppm
BURTON
WATER CRYSTAL
To 'Burtonize' the liquor in malt extract beers
200g
How to Burtonize Your Brewing Water
If you're brewing and IPA and you want a good, firm bitterness you'll want to Burtonize your brewing water, AKA make it like the water in Burton-on-Trent where IPAs were made initially. The water in Burton-on-Trent is very hard. To match it exactly, the ions in your water should have the following profile:
Calcium (Ca+2) 352.0 ppm
LD Carlson
COMPANY
BURTON WATER SALTS
ADD 1 PKG/5 GAL JUST PRIOR TO BOIL
CONTAINS: GYPSUM
SALT130
BREW-RITE
DUBLIN
Water additive salts
The Grape and Granary
Edinburgh Salts
SALT130

SAMUEL SMITH
INDIA ALE
Smith's
OLD BREWERY
PALE ALE
SAMUEL SMITHS
INDIA PALE ALE
INDIA ALE
OLD BREWERY
TADCASTER
TADDY
BITTER
TADDY
GOLDEN ALE
PALE ALE
Samuel Smith
Triassic red sandstone
Upper Permian marl
Upper Magnesian Limestone
Middle Permian marl
Upper Permian marl
Lower Magnesian Limestone
Tadcaster
Magnesian Limestone
(blues)

to water. Such conditions are provided by the deep soils overlying the sandstones of Wolnzach, in Germany's Hallertau, now the pre-eminent hop area in Europe.

Hop growing in the British Isles long ago withdrew to Kent in southeast England, and to Herefordshire, partly because of the suitably well-drained soils derived from the underlying Old and New Red Sandstones (*see* Chapter 16). And as with wine, bedrock geology influences the physiography of the land. Hops are very prone to mildew, one reason why relatively dry Kent has remained important. The Herefordshire–Worcestershire area lies in the rain shadow of the Welsh Hills to the west, composed of older, much tougher rocks. In the USA, much of the hop crop comes from Washington and Oregon, from deep alluvial soils in the lee of the nearby Cascade volcanic uplands.

The three main steps in brewing are steeping the cereals in warm water (the 'mashing'), boiling the resulting liquid with hops, and then fermenting with yeast. The chemistry of the water influences all three stages, but especially the mashing. There the grains are converted into fermentable forms, their nature depending mainly on the temperature and pH of the mixture. Temperature has long been easily measured and controlled, but the matter of acidity always used to be clothed in mystery. Today, however, there is a vast literature on brewing chemistry, and the role of water in governing pH is well understood. (And brewing beer uses an awful lot of water. According to the World Water Organization, from growing the barley through to getting the finished beer in your hand takes no less than an astonishing 300 litres of water – per glass.)

Ideally, for efficient mashing the water should have a pH of around 5–5.5, so historically, hard, high pH waters were undesirable. To the rescue, however, came roasted malts. Toasting the malt encourages the natural phosphates to form mild phosphoric acid, which decreases the pH into an acceptable range. The effect is greater with higher degrees of roast but, of course, the beer will be darker.

We now know that calcium and magnesium are the key minerals in the water influencing brewing methodology, and that it's carbonate, bicarbonate and sulphate that affect the flavour. And here's the thing. Many breweries still today get their water from private wells, and from these the chemistry will very much depend on the geological nature of the aquifer. There's a plethora of technical

OPPOSITE TOP: A sample of the additives available to home brewers for simulating the waters of some classic brewing cities. BOTTOM: Classic English ales are produced at Tadcaster, England, because of the waters coming from the Permian-age marls on which the town is situated.

detail on how to adjust the water chemistry for brewing, but there is rarely any acknowledgement of the more fundamental influence, that the starting water depends on the local geology. I have over a dozen books on beer and brewing, some with entire chapters devoted to water, and in only one of them is geology even mentioned. Yet the aquifer geology was paramount historically; here's why.

In the UK, no brewing centre is more famous than Burton-on-Trent, in the English Midlands. It was unclear until around 1880 or so why the pale ale produced there was so good, though the suspicion always was that the explanation somehow lay in the water supply. It turns out that the water at Burton-on-Trent contains significant magnesium together with a remarkably high sulphate content. (Magnesium sulphate, Epsom salts, was once celebrated for its purgative properties, with people at one stage drinking up to 16 pints a day of Epsom water – with 'various funny results').

At Burton, however, the magnesium is balanced by high calcium and this allows high hopping rates. In short, Burton water has a perfect balance for brewing hoppy pale ale, together with levels of sulphate that give an unrivalled taste and bite (still known to aficionados as the 'Burton snatch'). The high hopping rates enhance the stability of the beer, so important in those days before pasteurization. And this enabled Burton beers to withstand being transported long distances.

Indeed, if the grain and hop content was increased, the resulting strong beer was stable enough to be exported away from the UK – the origin of 'Export Ale'. And if the ale was brewed yet stronger it could even survive the long sailing-ship journey to the colonies in India – the birth of today's almost ubiquitous India Pale Ale or IPA.

Nowadays this chemistry knowledge is standard in brewing textbooks around the world. So if the water supply of some modern brewery is inappropriate for brewing pale ale then it is simply adjusted to Burton levels, usually by adding calcium and magnesium sulphates – what are commonly called Burton salts – that is, by 'burtonizing' the water! However, much less often explained is why Burton has such water, and the answer, of course, lies in its geology.

The bedrock at Burton is a mudstone and underlying that is the Sherwood sandstone aquifer, both part of the New Red Sandstone formation. The mudstone formed in the desert conditions of Triassic times, which caused the mineral gypsum, calcium sulphate to form, and this is the key. The water from the aquifer has to pass through all that gypsum. For example, just the uppermost 100 metres of the well sunk for Burton's Worthington brewery encountered 22 metres of gypsum-bearing beds, some of which approached four metres in thickness.

Similar hydrogeological conditions existed at other early brewery towns. Yorkshire's Tadcaster, for example, has Permian-age marls containing seams of gypsum. Hence it is no accident that this small country town still boasts three major breweries, including the sainted Samuel Smith's. In fact, when that company recently considered selling their brewery well-water in bottles, they had to abort the plan because the high level of calcium sulphate was outside the parameters legally allowed in 'spring water'.

Over on the continent, parallel developments were taking place, especially in the German-speaking Austro-Hungarian Empire. New strains of sharp, aromatic hops were emerging, together with so-called 'bottom fermenting' yeasts that operated best at lower temperatures. Newly invented refrigeration made this possible, together with cool storing of the finished product, a process known in German as 'lagering'. And just as in England, one town set the benchmark for this new, clean tasting *lagerbier* – Plzeň in the Czech Republic. In German it was known as Pilsen, and its product the pilsener, a name found today in various forms of spelling on beer labels around the world.

In contrast to Burton water and its ales, this more delicate style of beer is best produced from soft, low mineral content water – just as at Plzeň. There the water is drawn from Carboniferous-age sandstones and ancient metamorphic rocks that give minimal interaction between the aquifer and the water. Consequently, the water has a remarkably low mineral content. Moreover, the massive, softish sandstones beneath Plzeň allowed vast systems of underground caves to be hewn, ideal for the cool lagering of the pilsener beer.

The water at České Budějovice, 150 kilometres (90 miles) to the southeast of Plzeň, is even purer, from a 300-metre deep (980-foot) well in old (Precambrian) gneisses. German speakers know the town as Budweis and its beer as the Budweiser. The original Budweiser beer is still produced there. Other European brewing centres in general have less pure waters, for a variety of geological reasons. For example, around 85 percent of Brussels water comes from Eocene-age marine sandstones, some of which are calcareous, giving a calcium content that allows the production of some of the formidable range of beer styles for which Belgium is renowned. Recent siliceous sands and gravels with some marine, calcareous input give calcium-bearing moderately hard waters in both Copenhagen and Haarlem.

'Brewed with pure Rocky Mountain spring water' is a brilliant marketing phrase, but almost all Colorado breweries use municipal surface water rather than springs, and some of it is moderately hard. Denver water, for example, comes

from river catchments that in their lower reaches are underlain by Jurassic and sedimentary strata that are distinctly calcareous in parts. Water for the Coors brewery in Golden, Colorado, is also surficial, and Fort Collins breweries such as the New Belgium Brewing Company, Anheuser Busch and Odell's also use river and reservoir water. Fortunately, the mineral content is sufficiently low as to allow manipulation for a wide range of beer styles.

But what about those brewing centres with waters that lack a suitable natural balance of minerals? Munich breweries, for example, near the famous hop fields, chiefly draw their water from sands and gravels that, deriving from the nearby calcareous rocks of the Bavarian Alps, give a high carbonate content and high alkalinity. Not good for mashing, but here's where the roasted grains come in, which is why Munich has become best known for its darker beers. Even the paler lagers produced today have a richer taste than, say, pilseners, partly because of the unfermented compounds induced by the relatively hard water.

Vienna waters, although also derived from calcareous rocks in the Alps, are less hard and therefore require a lesser proportion of roasted malt. The result is the characteristic red-amber Viennese lager. Similarly, hard waters in London from the chalk lead to copper-coloured ales, and, where greater degrees of malt roasting were needed, the historically famous dark ruby porters of the city.

But perhaps the extreme case of using roasted, high-phosphate malts to reduce the pH levels of mashing water to acceptable levels is in Dublin, and most famously

ABOVE: The Pilsner Urquell Brewery in Plzeň, Czech Republic, which in 1842 produced the world's first pale lager, giving rise to the beer style known as pils or pilsner.

at the Guinness brewery. The building is sited on bedrock of Carboniferous limestone and obtains its water from a reservoir in the Wicklow Mountains to the west. It's fed from limestone and limestone-derived gravels that make the water distinctly hard and alkaline. But highly roasted malt – known as black malt – lowers the mash pH adequately. The resulting liquor leads to a beer high in tannins, with a grainy mouthfeel, and the very dark colour characteristic of Dublin or Irish stout. With such an idiosyncratic beer, in the early days Arthur Guinness had to devise ingenious advertising to promote it; today it's one of the world's best loved and most distinctive drinks – and essentially it owes its character to geology.

The 20th century saw great changes in the brewing scene, especially in the USA and the UK. Among other things, local products almost disappeared as beer was increasingly fabricated in a seriously reduced number of breweries. Regional identities became blurred or lost. Geology became largely irrelevant.

But in the 1970s America's youth was restless, and through increased travelling abroad became aware of Europe's brewing heritage. And in 1978, around half a century after Prohibition, home brewing again became legal and there followed the explosion of the craft beer/microbrewery movement. The new American brewers showed huge innovation but also had respect for the classic styles of beer – they rescued IPA from near extinction to become the mainstay it is today.

And motivated by this upheaval in the States, and Gordon Brown's 2002 budget tax-breaks for small brewers, an equivalent movement erupted in the UK, taking inspiration from the new, hop-driven American styles of beer. The number of UK breweries grew from 403 in 2002 to well over 3,000 in 2020, with a bewildering range of beer styles. (We should, however, keep all this new upsurge in volume in context. The merger, in 2016, between brewing colossi Anheuser-Busch InBev and SABMiller means that this single company produces almost a third of *all the beer drunk in the world.*)

Nowadays the water in breweries of whatever size is usually artificially adjusted to suit the style of beer to be produced, but even so the local geology provides the starting point. And now some small-scale brewers are making a concerted effort to create beers that reflect their locale – there is much talk of beer terroir. So besides using local water, these innovators are using wild yeasts, barley grown locally, and flavourings such as bog myrtle, woodruff, heather and yarrow. And there is some evidence that, as with grapevines, these plants depend on the soils in which they grow. Beer has a long and proud history, but that unsung role of geology – could it be that it's about to change?

Epilogue:

Can you really taste the limestone, smell the slate?

The massive granite of South Dakota's Mount Rushmore and Rome's monuments of tuff (or should that be tufa?), London's limestone memorials and the dark slate roofs of Angers, *La Ville Noire*: we've visited them all in the preceding chapters during our rambles through the wine world. We've seen in New Zealand the astonishingly stony but now coveted gravels that once belonged to William Gimblett, and in Australia the fiery-red terra rossa of Coonawarra that positively glows at sunset.

We've been to Dansaert in Brussels where the story of discovering photosynthesis began, and Paradise Valley in California, momentous for our modern understanding of earthquakes. We've met the almost forgotten Nicholas Pettinger, who devised a clever depiction of the pH-nutrient availability relationship that's now in just about every work on practical agriculture, and Pierre Viala, who in the Texas Hill Country located wild vines that resisted phylloxera yet thrived on limestone, thus ending the plight of French *viticulteurs*.

They are all features of the rich tapestry that is the world of wine. And like a thread running through the weave, in the preceding chapters I've kept pondering on what all this vineyard geology might mean for wine. As we saw in Chapter 2, for long it was thought absolutely fundamental, its pivotal role beyond question – until we discovered photosynthesis. After that, for a long time, geology took a back seat; books hardly mentioned it.

I, myself, first became interested in wine as an undergraduate student in the 1960s and started buying books. I still have them. I've just checked and Cyril Ray, Alec Waugh, Alexis Bespaloff and Michael Broadbent, to take four examples, barely mention rocks or soil at all. Moreover, several academic studies around that time failed to find much correlation between geology and wine, apart from where it became associated with water provision.

Even so, the notion of a link with the soil never really went away and late last century it burst back into prominence. At that time, wine was greatly improving in quality and increasing amounts were being consumed: competition was intensifying. Producers had to strive for better fruit.

As we have seen (in Chapter 1), the vineyard soil influences grape quality, and, indirectly, a wine's character. So growers now had to get to know their soils and make sure their vines and viticultural methods were optimally matched. Remote sensing of soil properties using satellite imagery and then drones was becoming more easily practicable and soon vineyards became peppered with exploratory soil pits. These methods are all now the norm and rightly so. And, as James Busby had noted nearly 200 years earlier (*see* Chapter 14), because the geology of a vineyard is hard to emulate elsewhere, producers eagerly started promoting some specialness of their soils. Geology was back!

But what about the wine in our glasses and its taste? Of course, it's now commonplace to relate taste to the soil, indeed some commentators claim that individual rock types imbue a wine with particular traits. There's talk of the 'tells' of soil types and 'soil signatures'. The idea is that the geological nature of the soil shows through in the finished wine, irrespective of how it was made or where it is from. And this where I have problems.

Soils derived from volcanic rocks are said to give 'fiery' wines, with 'pungency', 'spiciness', 'smokiness'. I wonder if those words would be used if the taster didn't know the soils were volcanic in origin. A wine from New Zealand's Gimblett Gravels (Chapter 12) has something called a 'gravelly tension'. Hmmm. I'm told that shale leads to wines that are 'elegant and complex', whereas slate gives 'rich, spicy and complex' wines and schist 'brings a seriousness to the wine'.

But we have seen (in Chapter 7) that shale, slate and schist are more or less the same chemically, as, for that matter, is granite. All normally provide good drainage and adequate water, so I find it hard to see how each of these rock types leads to wines with different tastes. In any case, the claims are noticeably inconsistent, as we saw in Chapter 4 regarding granite.

As far as I can see, journalistic 'blind' tasting sessions to assess such claims are usually inconclusive and it's formidably difficult (and there's little incentive) to design rigorous scientific tests. I did mention in Chapter 11, though, a recent controlled experiment in which trained tasters were able to recognize wines from different cultivars and fermented using different yeasts, but unable to detect any differences between the same wines originating from different soils.

One part of my problem is that the taste claims are never accompanied by any explanation of how they might come about. At best, there is some loose allusion to the vine's nutrient uptake. However, I outlined in Chapters 1 and 3 the various 'disconnects' between the chemistry of vineyard soils and the nutrient minerals that end up in the wine, and I noted in Chapter 15 that in practice most of them come from the soil humus anyway.

Despite all that, in what seem the most extreme claims, some commentators maintain they can actually taste a rock in the wine – 'you can taste the granite', 'distinct whiffs of basalt', 'the volcanic ash is right there', that kind of thing. However, as we saw in Chapters 1 and 3, pieces of rock, being insoluble, cannot be taken in by vine roots, let alone be transmitted through to the finished wine (where presumably we would see little specks of them).

And this insolubility means that these geological materials and rocks have no taste, and their inability to vaporize means they have no smell. Thus, although words such as 'mineral', 'chalk' and 'graphite' sound earthy, as tasting terms they cannot be literal. *They have to be metaphors* (Chapter 3).

Fashions come and go in the wine world. We talk about wine in changing ways. Would anyone today write that Bordeaux wine 'in its scansion ranges from the Horatian to the Miltonic, from the rippling lyrics of Herrick to the sway and surge of Swinburne', or that Sauternes is 'delightful in the right company, which is a plump, pretty and rather greedy young woman'?

I remember the era of 'flying winemakers', when the vineyard was deemed of relatively little relevance. And there was a craze for oak. Exactly how was the oak ageing carried out? What kind of oak? From which forest, where in the forest; where in the tree? Yet today, 'great wine is made in the vineyard' is the mantra, and it's the paucity or complete lack of oak that is lauded.

Might wine-tasting-of-soil go the same way? I don't know. It might fade to be replaced by some new fashion, but on the other hand maybe we'll discover that soil affects wine in ways not yet recognized and it will take on a new importance. Scientific understanding is always evolving, in line with new evidence. But as for the question of directly tasting a rock in wine – whether we really can 'taste the limestone, smell the slate' – the answer surely has to be no. And it's hard to see how that can change because it's so readily demonstrated (by licking and smelling a *clean, fresh* surface of the rocks). That evidence seems incontrovertible.

Anyway, the world of wine moves on, continuing to evolve in a host of different ways. I am seeing it get bigger, with new threads in the tapestry. From the dominance in the past of perhaps half a dozen grape types, indigenous and almost forgotten varieties are being enthusiastically rescued. Ancient fermentation vessels and arcane methods are being resuscitated. Countries that once were unimaginable as wine producers are now capitalizing on new know-how and climate change to throw their hat into the ring. And long-established countries are extending vineyards into previously unthinkable areas.

One Chilean producer (Ventisquero) has for years produced wines from the well-known regions in the central part of the country; now in Patagonia it has the world's most southerly vineyard, battling winds of 100 kilometres (60 miles) an hour, and another up in the Atacama Desert, the driest place on Earth. The latitude difference between those two sites is around 18 degrees: that's like a proprietor in Bordeaux also producing from vineyards in Sweden and Algeria. In other words, even if wine doesn't really taste of rocks, there are plenty of new horizons for a wandering geologist.

About the author

Alex Maltman is Emeritus Professor of Earth Sciences at Aberystwyth University in Wales. Alex has a long and decorated career in teaching and in academic research, which has involved some of the Earth's high mountains, drilling below the floors of the deepest parts of the oceans, and a lot of geological matters in between. And during all this time, his fascination with wine was growing. For 50 years Alex has himself grown vines and made wine at his home in Wales. It's strictly a hobby and there are only 20 vines – but then the basic principles and the decisions are just the same as dealing with 20,000 vines! (Though with rather less pressure.)

Right from the start of this viticultural odyssey, Alex's scientific curiosity questioned why things were done in certain ways and, inevitably, the much mentioned but poorly understood relationship between wine and vineyard geology. This led to numerous publications on the subject, in both the popular press and academic journals.

Alex has researched and lectured on vineyard geology on four continents, and is a seasoned speaker at international conferences. He has contributed to a number of wine books, including the *The World Atlas of Wine* and *The Oxford Companion to Wine*, for which he is responsible for the geological entries. Alex is the author of the acclaimed book *Vineyards, Rocks and Soils: A Wine Lover's Guide to Geology* (Oxford University Press, 2018).

Index

Page numbers in *italic* refer to the illustrations

Acknowledgements

Trying to recall and acknowledge all the people who have brought about this book and inspired its content is like trying to count the stars in a clear night sky. Unfortunately, it's not feasible. But, rather like seeing the nearest planets, three lights stand out.

Neil Beckett, editor of *The World of Fine Wine* magazine, has always given me warm encouragement, from the first time I gingerly offered him a manuscript through to his graciously arranging permission for the articles he went on to oversee to be used in the present anthology. Hermione Ireland, managing director of the Académie du Vin Library, has given her energetic and enthusiastic backing right from the beginning of the project. Quite simply, without the dedicated support of Neil and Hermione, this book would not have been possible.

It has been a pleasure to work with Susan Keevil, Académie du Vin Library's editorial director. Susan's interest in the work was evident from the very start, in her patient but enthusiastic attentiveness in helping the book take form. Our interchanges have been numerous, constructive and convivial; in them Susan has dealt equally carefully with everything from the book's overall structure to the placing of a comma.

I must also thank Charles Curtis for writing such a warm Foreword, and James Pople for his stylish design work.

Only when the sun has set does the starry night sky appear; but, albeit temporarily out of sight, the sun is always there, always dependable. Similarly, my wife, Joanne, has been my unswerving support and companion not only while preparing this volume, but throughout the 50 years and more since I first put some writings into print.

Picture Credits

Photographs on the following pages are credited to the author, Alex Maltman:
18–19, 27, 28, 34, 36, 39, 51, 56, 60, 61, 62–63, 80–81, 86, 97, 102, 110, 112–113, 125, 131, 141, 150, 166, 169, 171, 174–175, 179, 184, 192–193, 196, 207, 230, 238, 244, 247, 248, 250, 276, 280, 283, 284, 288.

Other credits are as follows:
Page 4 Walkerssk
Page 42 AMHYDRO, Arcata, California
Page 45 Brent Spry, Madison, Indiana
Pages 76–77 Frédérique Hermine
Page 201 Sullivan Rutherford Estate
Page 262 Richard and Iola Huws, Pant Du vineyard
Pages 294–295 Roman Biernacki

The photographs on the following pages are used under license:
Adobe Stock
Pages 18–19 Kushnirov Avraham
Page 199 Barmalini
Alamy Stock Photo
Page 134 Stephane Gautier
Page 136 Zoonar/Nikki Gensert
Pages 154–155 Zoonar/Rvo233
Pages 208–209 Skyscan Photolibrary
Pages 256–257 Howard Taylor
Creative Commons CC0 1.0 Universal Public Domain Dedication
Pages 126–127 Hermann Hammer
Creative Commons license Attribution 2.0 Generic
Page 21 Woodloper/Woodwalker
Page 108–109 Ian Kluft
Page 142–143 Arby Reed
Page 72 South Africa Tourism
Page 217 James St John (cropped picture)
Creative Commons Attribution-Share Alike 3.0 Unported
Pages 270–271 Terry Joyce
Page 279 Ayack
Creative Commons Attribution-Share Alike 4.0 International
Pages 46–47 The Trustees of the Natural History Museum, London
Page 190–191 Florian Pépellin
Pages 212–213 Portable Antiquities Scheme
Page 227 Alexander Turnbull Library (above), R. Howe (below)
Getty Images
Pages 240–241 Kateryna Kon/Science Photo Library
Shutterstock
Page 163 firdausraka

The photographs on the following pages are used from Pexels.com:
12–13, 30–31, 94–95, 224–225.

The photographs on the following pages are used from Unsplash.com:
9, 67, 91, 158–159, 180–181, 272–273.

Académie du Vin Library and Professor Alex Maltman are grateful for the co-operation of Neil Beckett, Editor of *The World of Fine Wine* magazine, for his assistance in the publication of this book. Chapters 1 and 4–14 (inclusive) were originally commissioned for the magazine and are reprinted here with the kind permission of its publisher.

Estate Bottled
ALLUVIUM
AUGUSTA BLENDED RED WINE
di Breccia
AMPHIBOLITE
CALCAREOUS
VINEYARD
GREY
ANDESITE
Spring Mountain District
Napa Valley
CAMBRIAN
ARAGONITE
ANTHONY ROAD
DEVONIAN
Arkose
VQA NIAGARA ESCARPMENT VQA
Dolomite
2011
BASALT
ESCARPMENT
MARTINBOROUGH
TOSCANA
BEDROCK
Wine Co.
BIOTITE
GALENA
2014
Blue Slate
Riesling Kabinett
AUSLESE